Environmental Cost and Face of Agriculture in the Gulf Cooperation Council Countries

Shabbir A. Shahid • Mushtaque Ahmed
Editors

Environmental Cost and Face of Agriculture in the Gulf Cooperation Council Countries

Fostering Agriculture in the Context of Climate Change

Springer

Gulf Research Centre Cambridge
Knowledge for All

Editors
Shabbir A. Shahid
Soil Management Program, Research and Innovation Division
International Center for Biosaline Agriculture
Dubai, United Arab Emirates

Mushtaque Ahmed
Department of Soils, Water and Agricultural Engineering
College of Agricultural and Marine Sciences
Sultan Qaboos University
Oman

ISBN 978-3-319-05767-5 ISBN 978-3-319-05768-2 (eBook)
DOI 10.1007/978-3-319-05768-2
Springer Cham Heidelberg New York Dordrecht London

Library of Congress Control Number: 2014942259

Cover image caption: Protected agriculture is gaining importance in Gulf Cooperation Council (GCC) countries. This photo illustrates farmers making a gradual shift from open field agriculture to protected agriculture (Photograph courtesy of John A. Kelley, USDA-NRCS).

Printed on acid-free paper

Springer is part of Springer Science+Business Media (www.springer.com)

Foreword

The 2008 global food crisis and the sudden increase in commodity prices brought the issues of food security and sustainability of food production to the forefront in the Gulf region. Despite the issue having always lingered somewhere in the background, the fact that the GCC countries import more than 70 % of their food requirements was suddenly thrust forward as a high-priority national security issue with governments in the region scrambling to formulate national strategies and food security plans. Included in this flurry of activity sweeping the region were announcements of multi-billion dollar investments abroad in an attempt to ensure food supplies and provide for long-term food security. However, many of these investments have failed to come about due to a variety of political and economic factors. As such, food security remains a subject of concern for the region.

In the Gulf, water scarcity is pronounced and dependence on food imports continues to grow for which there are no easy solutions. Moreover, it must be understood that the issue of food security is closely intertwined with a host of other challenges confronting the Gulf region including climate change, reliance on hydrocarbon income, political instabilities, and rising population growth including the continued reliance on a large expatriate workforce. All these factors combine to make agriculture production in the region a topic of current relevance.

This volume is the result of a workshop entitled "Environmental Cost and Changing Face of Agriculture in the Gulf States" organized within the framework of the 2012 Gulf Research Meeting held from July 11–14, 2012 at the University of Cambridge. The workshop brought together participants from a wide variety of backgrounds and geographical locations in an effort to better understand the challenges the Gulf region faces when it comes to food production and how to bridge the gap between local production and food imports. The book highlights not only the prospects of agriculture in the region but looks at the potential of climate-smart agriculture, improved water use efficiency, the use of treated wastewater as well as issues of agricultural research and development.

The Gulf Research Centre Cambridge is grateful to Dr. Shabbir A. Shahid, Senior Scientist at the International Center for Biosaline Agriculture (ICBA) in Dubai, UAE, and Dr. Mushtaque Ahmed from Sultan Qaboos University in the

Sultanate of Oman for having taken on the task of directing this workshop and then following the work through to this publication. A special thanks is also extended to all those who presented papers at the workshop and thus added to our knowledge about this important topic for the Gulf region. It is through such work that we hope to enlighten the policy debate and contribute to further understanding some of the challenges faced by the region.

Chairman
Gulf Research Centre Cambridge

Dr. Abdulaziz O. Sager

Message

Worldwide the main issues hampering food security are water/land availability and the food distribution. The numbers are worrisome: 870 million people are hungry; every year about 12 million hectares worldwide are lost to land degradation, and the rate is increasing exponentially. Further, based on the World Economic Forum, water supply is among the top five global risks by likelihood and impact and the GCC is the most water-scarce region in the world.

Studies have shown that to increase the world production by 2050, we will have to increase agriculture production in marginal environments. And to do so, we need innovative technologies and ideas to use wisely all available resources. The International Center for Biosaline Agriculture has worked, over the last 14 years, closely with several GCC countries to increase local agriculture production and develop National agriculture and food security strategies. Over the last year, ICBA went through a participatory strategy review and development exercise (including a foresight symposium) which resulted in a new strategy emphasizing innovative research, partnership and impact. We, at ICBA, aim at working with partners to deliver agricultural and water scarcity solutions in marginal environments.

This book is a good illustration of joint efforts between Gulf Research Centre Cambridge, Sultan Qaboos University, and ICBA. I would like to congratulate Gulf Research Centre Cambridge for organizing the "Environmental Cost and Changing Face of Agriculture in the Gulf States" workshop and Drs. Shabbir A. Shahid and Mushtaque Ahmed for conducting the workshop and editing this important book.

Director General
International Center for Biosaline Agriculture
Dubai, United Arab Emirates

Ismahane Elouafi

نزرع للغد
ICBA
AGRICULTURE FOR TOMORROW

Message

I am pleased to know about the publication of this book highlighting the environmental costs and the changing face of agriculture in the GCC countries. Agriculture faces serious challenges in the GCC countries due to water shortage, arid condition, poor soil, climate change effects and various other reasons. Agriculture plays an important role in Oman as large number of Omanis are engaged in agricultural activities. Agriculture in Oman is also the largest consumer of water resources (more than 90 %). As such there is a need for research to ensure partial self-sufficiency of some crops and attain food security in Oman without harming the environment. The faculty and staff of the College of Agricultural and Marine Sciences of Sultan Qaboos University are engaged in research to tackle problems of Omani agriculture and have made enormous contribution in broadening the knowledge base on different issues relevant to GCC agriculture. This book will help in disseminating such knowledge to a wider audience.

I congratulate Gulf Research Centre Cambridge (GRCC) and the editors for publishing this book which will be of interest not only to the agricultural scientists but also to the decision makers of this region.

Deputy Vice-Chancellor – Postgraduate Studies and Research
Sultan Qaboos University
Muscat, Sultanate of Oman

Prof. Amer Ali Al-Rawas

Preface

Gulf Research Centre Cambridge (GRCC) organizes the annual Gulf Research Meeting (GRM) which seeks to provide an academic environment to foster Gulf studies and promote scholarly exchanges among scholars working on/or having familiarity with the Gulf region. GRM identifies subjects of importance to the Gulf region, stimulates research in these subjects, and provides a forum for broad dissemination of the research results. Motivated by the same objectives underlying the Gulf Research Meeting, the GRCC also serves as a platform for other events throughout the year and provides a focal point for students dedicated to carrying out and promoting critical research related to the Gulf.

This volume is the outcome of an Agriculture Workshop held in Cambridge University, UK, during the Gulf Research Meeting 11–14 July 2012. The GRCC brought together 450 distinguished scientists and policy makers from 46 countries to participate in 19 workshops. The Agriculture Workshop "Environmental Cost and Changing Face of Agriculture in the Gulf States" was attended by participants from Australia, Bahrain, India, Kuwait, Oman, Saudi Arabia, Turkey, UAE, UK, and Morocco. The agriculture workshop was directed by Dr. Shabbir A. Shahid, Senior Scientist at International Center for Biosaline Agriculture (ICBA) Dubai, UAE, and co-directed by Dr. Mushtaque Ahmed from Sultan Qaboos University, Sultanate of Oman.

The main objective of the agriculture workshop was to bring together scientists, educators, researchers, policy makers and managers from around the world who somehow have been involved in agriculture sector in the Gulf States to share their experience to improve agriculture production to bridge the gap between local production and the food import. GRCC received an overwhelming response to the call for the papers. These papers were reviewed and those deemed appropriate to workshop theme were accepted.

The papers contained in this book were presented at the workshop, and passed through rigorous peer review by renowned scientists. The papers are diversified and present various aspects of agriculture production in the changing face of climate change and dwindling water resources in the region. The book covers topics such as, prospects of agriculture in a changing climate, potential of climate smart agriculture, food security, improved water use efficiency, challenges in using treated wastewater, investment in foreign agriculture, and agricultural research and development.

At the conclusion of the agriculture workshop, following recommendations were made: (1) create "Gulf Agriculture Network for Knowledge Sharing and Technology Adoption" GANKSTA, (2) increase investment in research, development and extension in the agriculture sector, (3) adopt climate-smart technologies and practices in agricultural intensification for food security enhancement, (4) increase strategic groundwater reserve through recharge using reclaimed water, (5) invest in the use of reclaimed water in protected agriculture and enhance the social acceptance of this alternative by education and awareness campaigns, (6) educate consumers on the positive impacts of reducing meat-based food products in order to help reducing food imports; mitigating climate change and managing sustainably water resources, (7) adopt options for the Gulf States food security through the involvement of host countries small-holder farmers in the foreign land deals, (8) orient the Gulf investors either by lobbying or incentive to lease unutilized land within the framework of foreign land deals, (9) policy-makers need to be more fully cognizant of, and responsive to, the problems of current land deal strategies and the risks these create and the difficulties these strategies present in achieving reliable long-term food supplies, (10) policy-makers should research and establish alternative investment structures and mechanisms to achieve long term food security goals – such as reliable sources of staples from the agricultural sectors of less-developed countries, and (11) enhance and stimulate quality local food production and its competitiveness by adopting modern technologies such as soilless and hydroponic growing systems in order to maximize the market share for local products.

We wish to take this opportunity to express our sincere appreciation to the Gulf Research Centre Cambridge for hosting the Agriculture Workshop under the framework of GRM 2012 and for the publication of the book. Special thanks are due to all authors and co-authors and reviewers without whom this comprehensive book could not have been produced. We also owe our gratitude to Dr. Christian Koch, Director, Gulf Research Center Foundation (GRCF), Florian Weisweiler, Business Development Manager, GRCF, and Dr. Abdulaziz Sager, Chairman, Gulf Research Centre (GRC) for their professionalism and dedication to the entire GRM 2012 and special interest in the agriculture workshop. We are very optimistic that this book will be of interest to researchers, experts, and policy-makers in the field of agriculture.

Dubai, United Arab Emirates
Muscat, Oman

Shabbir A. Shahid
Mushtaque Ahmed
The Editors

Acknowledgements

I have been privileged to share the responsibility of directing the agriculture workshop and the editing of this book with Mushtaque Ahmed (co-workshop director) whose expertise and commitment sped up the editing process and certainly improved the quality of this book.

On behalf of the participants of the Agriculture Workshop, we would like to sincerely thank the organizers of the Gulf Research Meeting 2012 for providing the opportunity to conduct this workshop. Special thanks go to Dr. Christian Koch, Director, Gulf Research Center Foundation (GRCF), Florian Weisweiler, Business Development Manager, GRCF, and Dr. Abdulaziz Sager, Chairman, Gulf Research Centre (GRC) for coordination – from proposal submission, to conducting the workshop and publication of this book. The GRC staff deserves special thanks for their untiring efforts in organizing this significant event in Cambridge.

Dubai, United Arab Emirates

Shabbir A. Shahid
Editor

About the Editors

Dr. Shabbir A. Shahid, a Senior Scientist at Dubai based International Center for Biosaline Agriculture (ICBA), earned Ph.D. degree from the University of Wales, Bangor, UK, in 1989. He joined ICBA in 2004 and has over 33 years experience (Pakistan, UK, Australia, Kuwait and UAE) in agriculture related RD & E activities. He has held several positions: Associate Professor (University of Agriculture Faisalabad, Pakistan), Associate Research Scientist in Kuwait Institute for Scientific Research, Kuwait, and Manager of Soil Resources Department, Environment Agency – Abu Dhabi, UAE. During his professional career he has been working in applied agriculture projects in many countries (Pakistan, Kuwait, UAE, Qatar, Oman, Niger, Morocco, Jordan, Spain, Syria etc.). In the GCC countries, he has been working for the last 18 years and is very much familiar with agriculture activities and food security issues. He is a prolific author of over 150 publications in peer-reviewed journals, proceedings, books and manuals. He is currently life member and councilor of the World Association of Soil and Water Conservation, member advisory board "World Forum on Climate Change, Agriculture and Food Security–WFCCAFS". Dr. Shahid enjoyed working as member of Scientific Committees, International Advisory Boards of International Conferences, and editor of scientific books, and reviewer of scientific journals.

Dr. Mushtaque Ahmed obtained his Ph.D. from Iowa State University, USA, in 1988. He is an Associate Professor of the Department of Soils, Water and Agricultural Engineering of Sultan Qaboos University (SQU), Oman. He also served as the Director of Center for Environmental Studies and Research (CESAR) of SQU from 2010 to 2012. He joined SQU in 1996. Prior to that he worked for various organizations in Australia (Land and Water Conservation Department–NSW; CSIRO–Perth; Department of Water Resources–Griffith, NSW). His current research interests are biosaline agriculture, managed aquifer recharge, climate change and adaptability, etc. He has published more than 140 scientific papers in peer-reviewed journals, book chapters, conference proceedings, and manuals, as well as serving as editor of various publications. He is well experienced in organizing workshops, conferences and symposiums at national and international levels. He is a corporate member of the Institution of Engineers, Australia, and a member of the International Association of Hydrological Sciences and the Asia Oceania Geosciences Society.

Contents

Contributors

Mushtaque Ahmed Department of Soils, Water and Agricultural Engineering, College of Agricultural and Marine Sciences, Sultan Qaboos University, Oman

Asadullah Al Ajmi Soils and Irrigation, Desert and Arid Zones Sciences Programme, College of Graduate Studies, Arabian Gulf University, Manama, Bahrain

Salwa Abdullah Nasser Al Jabri Planning and Investment Promotion, Ministry of Agriculture and Fisheries Wealth, Muscat, Oman

Saif Ali Alkhamisi Head of Field Crops Research, Ministry of Agriculture and Fisheries, Muscat, Oman

Habibah S. Al-Menaie Environment and Life Science Center, Kuwait Institute for Scientific Research, Safat, Kuwait

Salma Saeed Ahmed Bani Planning and Municipal Council Department, Ministry of Municipality and Urban Planning, Manama, Bahrain

Mohamed Behnassi Global Environmental and Health Politics, Faculty of Law, Economics and Social Sciences, Ibn Zohr University of Agadir (Morocco), Agadir, Morocco

North-South Center for Social Sciences (NRCS), Agadir, Morocco

Houcine Boughanmi Department of Natural Resource Economics, College of Agricultural and Marine Sciences, Sultan Qaboos University (SQU), Muscat, Oman

Mohamed Boussaid National Institute of Agronomic Research (INRA), Rabat, Morocco

R. Gopichandran Environment and Climate Change Wing, Gujarat Energy Research and Management Institute, Gandhinagar, Gujarat, India

Nejib Guizani Department of Food Science and Nutrition, College of Agricultural and Marine Sciences, Sultan Qaboos University (SQU), Muscat, Oman

Huzur Keskin Finance, International Investment and Affiliates Manager, Ankara, Turkey

Mumtaz Khan Department of Crop Sciences, College of Agricultural and Marine Sciences, Sultan Qaboos University, Muscat, Oman

Hemesiri Kotagama Department of Natural Resource Economics, College of Agricultural and Marine Sciences, Sultan Qaboos University (SQU), Muscat, Oman

Msafiri Daudi Mbaga Department of Natural Resource Economics, College of Agricultural and Marine Sciences, Sultan Qaboos University, Muscat, Oman

Sanmugam A. Prathapar Theme Leader, International Water Management Institute, New Delhi, India

Shabbir A. Shahid Soil Management Program, Research and Innovation Division, International Center for Biosaline Agriculture, Dubai, United Arab Emirates

Benjamin Shepherd Centre for International Security Studies, University of Sydney, Sydney, NSW, Australia

About the Authors

Dr. Mushtaque Ahmed obtained his Ph.D. from Iowa State University, USA, in 1988. He is currently the Director of Center for Environmental Studies and Research (CESAR) as well as an Associate Professor of the Department of Soils, Water and Agricultural Engineering of Sultan Qaboos University (SQU), Oman. He joined SQU in 1996 and was honored as outstanding teacher in the college of agriculture SQU Oman in 1999. Prior to that, he worked for various organizations in Australia (Land and Water Conservation Department–NSW; CSIRO–Perth; Department of Water Resources–Griffith, NSW). His current research interests are biosaline agriculture, managed aquifer research, climate change and adaptability, etc. He published scientific papers in peer-reviewed journals, book chapters, conference proceedings, and manuals, as well as editor of various publications. He is well experienced in organizing workshops, conferences and symposiums at national and international levels.

Dr. Asadullah Al Ajmi earned his Ph.D. degree in Soils and Irrigation Sciences from the University of Reading, UK. He worked at the College of Graduate Studies, Arabian Gulf University in Bahrain as an Associate Professor of soils and irrigation. Dr. Al Ajmi has over 25 years of experience in agricultural research design, implementation and analyses. He has been the Director for Soils and Water Research program at the Ministry of Agriculture in Oman for many years. He also served ICARDA (International Centre for Agricultural Research in Dry Areas) as an irrigation management expert. Dr. Al Ajmi has been actively involved in the preparation of various UN reports related to environmental issues including the GEO 4 (Global Environmental Outlook) report. Dr. Al Ajmi has published over 10 papers in international refereed journals, 6 in conference proceedings and several technical reports and concept notes.

Ms. Salwa Abdullah Nasser Al Jabri obtained B.Sc. and M.Sc. degrees in Natural Resource Economics from the Sultan Qaboos University, Sultanate of Oman. Ms. Salwa researched on household food security in the Sultanate of Oman for her masters studies. She currently serves the Ministry of Agriculture and Fisheries Wealth of the Sultanate of Oman as an Investment Analyst.

Saif Ali Alkhamisi, since 2006, is working as Head of field crops research in the Ministry of Agriculture & Fisheries, Directorate General of Agricultural & Livestock Research, Muscat Oman. He completed his master degree in Bioresources and Agricultural Engineering and is currently pursuing studies leading to Ph.D. degree program in Soil and Water Management at Sultan Qaboos University Oman. He published 21 papers in peer-reviewed national and international journals, a book on Principles of Statistics and Experimental Design, and a book chapter. He attended over 48 training courses, conferences, workshops and professional programs, and presented 14 papers and posters. He contributed to 16 technical reports related to field crops research and the reuse of treated wastewater in agriculture. He contributed five short articles in the local newspapers.

Habibah S. Al-Menaie, has secured her Ph.D. from University of Reading, UK, in Crop Production and Breeding with distinction. As a researcher, at KISR she has made significant contributions in strengthening research and development activities in Environment and Life Sciences Research Centre. She has extensive experience in greenhouse production of ornamental plants, integrated management in crop production, plant tissue culture, crop improvement of many plants including ornamentals, forage and sustainable agriculture. She is experienced in crop production under stress and harsh conditions. She is a Project Leader and Principal Investigator of many ongoing and completed projects focusing on adaptation and production of plants under harsh environmental conditions mainly olive, rose, forage, water plants flowering plants and argan. She has many awards to her credit, including KISR Scientific Achievement Award and is the first among the Kuwait Scientists. Also she received KFAS Scientific Prize for Biological Sciences during 2011. She has membership in different committee such as the Co-chair of second Arab-American Frontiers Symposium, 2013, Organizing Committee of International Conference on Women in Science and Technology in the Arab Countries, 2013, Co-Chair of the First Annual Symposium of the Arab American Frontiers of Science held at KISR from 17–19 October 2011, Kuwait, Co-Chairperson of Organizing Committee – Executive Board Meeting of Organization for Women in Science for the Developing World (OWSDW) May 30–June 1, and KISR Scholarship Committee, Grievance Cell and Strategic Program Review Committee. She is the author of three books, and has published many research scientific papers in refereed journals and scientific conferences and presented her scientific accomplishments in many regional and international conferences. She has submitted many annual research and technical/progress/final reports. Dr. Habibah has conducted a number of training courses for KISR staff and for participants outside KISR and has attended local and international training courses on different specialized areas.

Salma Saeed Ahmed Bani worked in diverse positions including agricultural economist, food policy specialist, head of commodity analysis unit with FAO and head of monitoring and evaluation section with GTZ project. Currently Ms. Bani is employed as a senior executive, strategic planning in the Ministry of Municipality and Urban Planning Kingdom of Bahrain. In the field of research, she carried out scientific research for many years, both in academics and as a researcher as a result of which she merited recognition award of best researcher on a study entitled "Option for Ensuring Sustainable Food Security in Bahrain" in 2010 under the program of "The Award of the Minister of Municipalities Affairs and Agriculture for Innovative and Agricultural activities". She has a track record of 20 years work and research experience, first as a research assistant with Salford university in UK, and then as a research scholar in Assessing the Competitiveness and Sustainability of Bahrain food production system: A policy analysis Matrix Approach. Bani is currently pursuing her Ph.D. degree in Sustainable Food Production System in Open University Malaysia. She earned her M.Sc. in International Agricultural Marketing – New Castle University, UK, 1990–1991; Diploma in Agricultural Marketing – New Castle University, UK, 1990–1991; and B.Sc. (Hons.) Agriculture – Somali National University, Somali, 1982–1985. Bani has published and presented papers in International Conferences in the Middle East and Europe.

Mohamed Behnassi, Ph.D. in International Environmental Law and Politics, is an Associate Professor at the Faculty of Law, Economics and Social Sciences of Agadir, Morocco. He is the Founder and Director of the North-South Center for Social Sciences (NRCS) and member of the advisory board of the World Forum on Climate Change, Agriculture and Food Security. His core teaching, research and expertise areas cover the Global Environmental and Human Security Politics, Corporate Social Responsibility, and Adaptation to Environmental Change. He has published many books with international publishers (such as: *Global Food Insecurity*, Springer 2011; *Sustainable Agricultural Development*, Springer 2011; *Health, Environment and Development*, European University Editions 2011; *Climate Change, Energy Crisis and Food Security*, Ottawa University Press 2011) and published numerous papers in accredited journals and communicated several oral presentations in relevant international conferences. In addition, Dr. Behnassi has organized many outstanding international conferences covering the above research areas. He had also completed in 2011 a US State Department-sponsored Civic Education and Leadership Fellowship at Syracuse University and the Maxwell School of Citizenship and Public Affairs (Syracuse, New York). Currently, he is the manager of a key research project on behalf of the Royal Institute of Strategic Studies (IRES) in Morocco titled: "Food and Health Security Face to Climate Change in Morocco: An Adaptation Strategy within the Perspective of an Integrated Governance".

Houcine Boughanmi is an Associate Professor in Agricultural Economics and Head of the Department of Natural Resource Economics at Sultan Qaboos University (SQU). He got his bachelor degree from the University of Tunis, then his M.Sc. from the University of Kentucky, USA, and his Ph.D. from Oregon State University, USA. Dr. Boughanmi has 25 years of experience in teaching, research and community services, and has published extensively in reputable academic journals. Dr. Boughanmi's research interest includes international trade using partial and general equilibrium modeling, food demand, food policy and food security. He is currently leading a research project on the effect of the food crisis on household welfare and contributing to another one on estimating consumer preferences for organic food using conjoint analysis. He also served as a co-investigator in a number of SQU funded research projects in the areas of agricultural production and farming systems in Oman. Dr. Boughanmi has recently organized an international conference to discuss food security strategies in the Arab Countries with key note speakers from IFPRI, the Borlaug Institute, and The Arab Organization of Agricultural Development and other international organizations.

Mohamed Boussaid holds a master's degree in agronomy (National School of Agriculture, Morocco). His focus is on food security and value chain development. In 2003, he joined the German Technical Cooperation (GIZ) as a senior technical advisor on environmental issues. In the last 7 years, his focus was on natural resources management, biodiversity, climate change and desertification control.

R. Gopichandran serves as the Director of Vigyan Prasar, an autonomous institution under the Department of Science and Technology, Government of India. Gopi's two doctoral degrees are in the areas of microbial and chemical ecology. His extensive work on preventive environmental management motivated him to secure a degree in law. Most of his 23 years of scientific, research, communication and capacity building work and output relate to chemical ecological aspects of natural resources management, cleaner production and chemical substitution. The latter two aspects have been through a dynamic association with the Compliance Assistance Programme of the United Nations Environment Programme in particular and with special reference to ozone layer protection, centred on locally relevant action by industry at the national and regional levels. This included innovative work on destruction of ozone depleting substances to twin the objectives of climate and ozone layer protection. His recent contributions to India's 12th Five Year Plan as a Member of the Working Group on Natural Resources Management are on the chemical ecological aspects of conservation and sustainable management of agriculture to tackle challenges posed by climate change. He has a large number of publications including books and papers through well recognized publishers and journals. He has also taught extensively in several centres of higher learning including the Centre for Environmental Planning and Technology and the Indian Institute of Management, Ahmedabad. Prior to joining Vigyan Prasar as its Director, he has served in three important institutions over the last 23 years. They are the Gujarat Energy Research and Management Institute, Gandhinagar, as the Principal Research Scientist/Scientist SH (Environment and Climate Change Wing); Centre for Environment Education, Ahmedabad, as Programme Director/Scientist SG – Environment Management and the Entomology Research Institute, Chennai. He is also an Alumnus (Year 2000) of the International Visitors Leadership Programme of the Department of State, USA.

Nejib Guizani obtained his Ph.D. in Food Science from the University of Florida, USA. After earning his Ph.D., he worked as Associate Professor in the School of Food Industry at the University of Tunis II. At present, he is working as Professor and head of the department of Food Science and Nutrition at Sultan Qaboos University, Oman. He conducts research based on a multidisciplinary approach combining food chemistry, processing, and microbiology. At early stages he has established a research program centered on the study of the properties and changes during processing and storage of indigenous foods. Food products studied include dairy products (Laban, cheese), dates and fish. Properties studied include physico-chemical, microbial, and state diagrams. Current research projects are in the area of functional foods. Specific interests include the antioxidant and anti-inflammatory effects of phenolics and flavonoids and related compounds using locally available edible species to enhance community health, provide income and conserve biodiversity. He has published more than 55 research articles in various peer-reviewed international journals, 23 conference proceedings, 18 abstracts and 9 book chapters. He has presented papers in 44 international conferences and 17 local seminars/workshops. He is associated with several international/national journals, as Associate Editor and editorial board member, and peer reviewer. He is a life member of Gamma Sigma Delta, The Honor Society of Agriculture, in recognition of his high scholarship, outstanding achievement and service to agricultural science.

Huzur Keskin is an executive consultant in Turkey, Balkans and UAE about M&A activities in the specific sectors which are energy, agriculture and husbandry, finance and international trade. Dr. Keskin has been giving lectures and seminars about the alternative financial sources at the universities. Dr. KESKIN was the Senior Project Director in Turkish Republic Prime Ministry Investment Support and Promotion Agency of Turkey (ISPAT). Prior to her appointment at ISPAT, she was a director in the corporate banking department of a public bank. Ms. Keskin has considerable experiences as a portfolio manager in the capital market at brokerage houses and as a financial analyzer in private banks. She holds a Ph.D. degree and a master's degree in "Foreign Direct Investment" and "Real Estate Investment Trusts-REITs" respectively. Keskin has published articles nationally as well as internationally, two of which are "Credit Derivatives Market" and "The Effects of FDI on Economical Growth".

Mumtaz Khan is serving currently as Associate Professor (Horticulture) at Sultan Qaboos University, Oman with teaching, research and extension responsibilities. He did his M.Sc. (Hons.) from the University of Agriculture, Faisalabad followed by a Ph.D. and postdoctoral training from the University of Sheffield, UK. Before joining the Department of Crop Sciences, SQU, he was serving the Institute of Horticultural Science, UAF, as faculty since 1986. At UAF, he founded Citrus Nursery Sanitation Lab. ever first in the country. He organized successful international symposiums and a national workshop sponsored by the external donors, and edited/published proceedings. At SQU, he took the initiative and started two major research activities on the use of waste-water and deficit irrigation in agriculture. He has been serving as PI and Co-PI in a couple of research projects at UAF&SQU. Dr. Khan has worked abroad for more than 10 years in some of the finest institutions, where he undertook diverse research and teaching assignments. Teamwork has been essence of his career as he demonstrated in executing joint research projects and publishing jointly in peer-reviewed journals and books. Dr. Khan served as head of research group on pomology and has held membership of several professional societies too. He has taken undergraduate/postgraduate courses in horticulture/agriculture and mentored large number of M.Sc. and Ph.D. candidates. He has published largely in peer-reviewed journals, extension articles, books and conference proceedings. In recognition of his services to science/ agriculture and the community, he was decorated with the prestigious Presidential Award during 2007.

Hemesiri Kotagama is an Assistant Professor at the Department of Natural Resource Economics, Sultan Qaboos University, Sultanate of Oman. He served at University of Peradeniya, Sri Lanka, as a teacher and Professor for 25 years. He has obtained his Ph.D. (Natural Resource Management/Economics), University of London; M.Sc. (Agricultural Economics), University of the Philippines at Los Banos; and B.Sc. (Agriculture) Honours, University of Peradeniya, Sri Lanka. He also has obtained diplomas/certificates on Environmental Economics, Gothenburg University, Sweden; Environmental Economics and Policy Analysis, Harvard University, USA; and Environmental Assessment and Management in Agricultural Development, University of London, UK. His research has been in the applied areas of environmental and natural resource economics and management. He has served in several national committees guiding environmental policy and in national technical committees evaluating major investment projects in Sri Lanka. He has been extensively involved in training professionals on environmental impact assessment and environmental economic valuation in the South Asian region. His current research is focused on food security and food waste.

Msafiri Daudi Mbaga is serving currently as Assistant Professor (Natural Resource Economics) at Sultan Qaboos University (SQU), Oman, with teaching, research and extension responsibilities. He did his M.Sc. (Agribusiness) from the University of Manitoba, Canada, followed by a Ph.D. (Economics) at the University of Manitoba Canada. Thereafter he was a Postdoctoral Research Fellow at Laval University, Quebec, Canada. Dr. Msafiri Mbaga also worked for the Canadian Government (Agriculture and Agri-Food Canada) in Ottawa for a number of years before joining the Department of Natural Resource Economics at SQU. Dr. Msafiri Mbaga's area of research include: demand analysis, food security, the effects of climate change on agriculture and adaptation and production economics.

Dr. Sanmugam A. Prathapar has an international reputation in agricultural water management and research management. During his 25 years of post-doctoral experience, he has been a researcher, academic, consultant and a development professional. He is now leading the research theme on "productive water use", at the International Water Management Institute and the Strategic Research Portfolio on Irrigation as a part of a CGIAR wide initiative on water, land and ecosystems. He has made over 200 technical publications and presentations during his career, which include over 50 peer-reviewed publications. Prior to joining IWMI, Dr. Prathapar was the Dean, College of Agricultural and Marine Sciences, Sultan Qaboos University, Oman; Professor at National Center for Groundwater Management, University of Technology, Sydney, Australia; and a Principal Research Scientist, CSIRO Division of Water Resources, Australia. He has held positions in Sri Lanka, USA, Australia, Pakistan, Oman and India, and has research and development experience in South Asia and Sub-Saharan Africa.

Dr. Shabbir A. Shahid, a Senior Scientist at Dubai based International Center for Biosaline Agriculture (ICBA), earned Ph.D. degree from the University of Wales, Bangor, UK, in 1989. He joined ICBA in 2004 and has over 33 years experience (Pakistan, UK, Australia, Kuwait and UAE) in agriculture related RD & E activities. He has held several positions: Associate Professor (University of Agriculture Faisalabad, Pakistan), Associate Research Scientist in Kuwait Institute for Scientific Research, Kuwait, and Manager of Soil Resources Department, Environment Agency – Abu Dhabi, UAE. During his professional career he has been working in applied agriculture projects in many countries (Pakistan, Kuwait, UAE, Qatar, Oman, Niger, Morocco, Jordan, Spain, Syria etc.). In the GCC countries, he has been working for the last 18 years and is very much familiar with agriculture activities and food security issues. He is a prolific author of over 150 publications in peer-reviewed journals, proceedings, books and manuals. He is currently life member and councilor of the World Association of Soil and Water Conservation, member advisory board "World Forum on Climate Change, Agriculture and Food Security–WFCCAFS". Dr. Shahid enjoyed working as member of Scientific Committees, International Advisory Boards of International Conferences, and editor of scientific books, and reviewer of scientific journals.

Benjamin Shepherd is completing his Ph.D. at the University of Sydney. Ben's research has examined the acquisition of agricultural land in developing countries by wealthier countries seeking to hedge against their own food scarcity risks. He has focussed on how these "land deals" are driven by state security concerns at the same time as creating new security and political dilemmas for the host states and affected populations. In 2011, Ben was a researcher in the Georgetown University School of Foreign Service in Qatar's Food Security in the Middle East research program. In 2010, Ben was Research Fellow with the Center for Non-Traditional Security Studies in Asia. In 2008, Ben was awarded the University of Sydney Hedley Bull prize in International Relations. Ben has a master's degree in International Studies and a bachelor of Industrial Design with Honours. Ben is currently working in the renewable energy sector. For 15 years, Ben was a senior corporate executive in the high-security technology industry. He also works in a Pro Bono capacity as an advisor to a charitable international aid organisation.

Acronyms and Abbreviations

AAGD	Arid Land Agriculture and Greenery Department
ACSAD	Arab Center for the Studies of Arid Zones and Dry Lands
AFED	Arab Fund for Environment Development
CCS	Climate Change Scenario
CGIAR	Consultative Group of International Agricultural Research
CSA	Climate Smart Agriculture
DI	Drip irrigation
EIA	Environmental Impact Assessment
FAO	Food and Agriculture Organization
GCC	Gulf Cooperation Council
GDP	Gross domestic product
GFN	Global Footprint Network
GHG	Green house gas
GRC	Gulf Research Centre
GRCC	Gulf Research Centre Cambridge
GRCF	Gulf Research Center Foundation
GSF	Gulf Salinity Forum
ICARDA	International Center for Agricultural Research in Dry Areas
ICBA	International Center for Biosaline Agriculture
IFAD	International Fund for Agricultural Development
IFPRI	International Food Policy Research Institute
IMF	International Monitory Fund
IPP	Intellectual Property Protection
IPR	Intellectual Property Rights
KISR	Kuwait Institute for Scientific Research
LAS	League of Arab States
LDCs	Least developed countries
MAF	Ministry of Agriculture and Fisheries
MCM	Million cubic meter
MENA	Middle East and North Africa
MoEW	Ministry of Environment and Water

MSL	Mean sea level
NAMA	National Appropriate Mitigation Actions
NAPA	National Action Plan for Adaptation
NAPLCD	Nutritionally Adequate Preferred Least Cost Diet
NRCS	North-South Center for Social Studies
R&D	Research and development
SDI	Subsurface drip irrigation
SLR	Sea level rise
SQCMSC	Sultan Qaboos Center for Modern and Soilless Culture
SQU	Sultan Qaboos University
STP	Sewage treatment plant
TDS	Total dissolved solids
TSE	Treated sewage effluent
TTSE	Tertiary treated sewage effluent
TWW	Treated waste water
UAE	United Arab Emirates
UN	United Nations
UNDP	United Nations Development Program
USEPA	United States Environmental Protection Agency
WB	World Bank
WHO	World Health Organization
WUE	Water use efficiency

List of Figures

List of Tables

Chapter 1
Changing Face of Agriculture in the Gulf Cooperation Council Countries

Shabbir A. Shahid and Mushtaque Ahmed

Abstract The combination of water scarcity, declining arable land, poor soils, hyper-arid environment and projected climate change impact in the Gulf Cooperation Council (GCC) countries (Bahrain, Kuwait, Oman, Qatar, Saudi Arabia and the United Arab Emirates) constrain the local economic agriculture production to meet food demand of the current and continuing population growth. This has raised serious concerns about food security in food insecure but capital rich countries, and thus food security has emerged as a goal to achieve in the GCC countries. There are various options to achieve food security, including but not limited to local agriculture intensification using technological innovations, food import, outsourcing food production in eco-creditor countries and leasing farmland abroad, etc. Currently over 70 % food is imported by the GCC countries, the reasons being the large deficit between the current biocapacity (resources generated) and the ecological footprint of consumption (resources used). Such a high food import risks food security when there are political instabilities, wars and famine in countries from where the food is being imported. It is, therefore, important to increase local production to reduce food import dependency. In doing so local agriculture production may be at the forefront. Many view agriculture in the region comes at some environmental costs as all the countries use much more water in agriculture than what is renewable. The challenge of balancing water demand against supply is enormous in the GCC countries. All the GCC countries are currently at various levels of water stresses. In this chapter the emphasis is given on various options to increase local agriculture

S.A. Shahid (✉)
Soil Management Program, Research and Innovation Division, International Center for Biosaline Agriculture, Dubai, United Arab Emirates
e-mail: s.shahid@biosaline.org.ae

M. Ahmed
Department of Soils, Water and Agricultural Engineering, College of Agricultural and Marine Sciences, Sultan Qaboos University, PO Box 34, Al-Khod 123, Oman
e-mail: ahmedm@squ.edu.om; ahmedm2316@gmail.com

S.A. Shahid and M. Ahmed (eds.), *Environmental Cost and Face of Agriculture in the Gulf Cooperation Council Countries: Fostering Agriculture in the Context of Climate Change*, DOI 10.1007/978-3-319-05768-2_1, © Gulf Research Centre Cambridge 2014

production and to examine their impact on the environment and to discuss alternate sound ways to achieve food security in the GCC countries.

Keywords Agriculture • Changing climate • Ecological footprint • GCCC • Strategic food reserve

1.1 Introduction

The Gulf Cooperation Council (GCC) countries are capital rich and have no foreign exchange limitation for food import. The region also benefits from the increase in oil and gas prices. Despite the government support to increase local agriculture production through financial assistance and subsidies, in 2008 agriculture input to Gross Domestic Product (GDP) was less than 1 % (Bahrain, Qatar, Kuwait), and 2.0 %, 2.8 % and 3.9 % in Oman, UAE and Saudi Arabia respectively. The increase in hydrocarbon sector resulted in lifting real GDP in the GCC countries by around 5.5 % in 2010–2011 compared with 5.2 % in early 2010.

In the Gulf States total bill of food import ballooning from US$ 8 billion to US$ 20 billion from 2002 to 2007 (Daniel 2011). According to estimates of the Economist Intelligence Unit, food import to the Gulf region, which stood at US$ 25.8 billion in 2010, are set to grow to US$ 53.1 billion by 2020. Due to water shortages and lack of arable land, the United Arab Emirates needs to import almost 90 % of its food requirements, UAE food imports are expected to mount to US$ 5.5 billion by 2015, rising to US$ 8.4 billion by 2020 (http://www.tradeandexportme.com/2014/02/whats-cooking/). Demand for food imports continues to be fuelled by significant demand from expatriates.

The collective nominal GDP of GCC countries was up by almost US$133 billion in 2010 and are expected to rise in 2011. The GCC countries nominal GDP is expected to hit US$ 1,010 billion in 2011. The population engaged in agriculture vary e.g., Qatar (0.8 %), Bahrain (1 %), Kuwait (1.1 %), UAE (3.2 %), Saudi Arabia (5.5 %) and Oman (29.2 %). In the GCC countries, annual population growth rate of 3.5 % was reported in 2008. Due to various constraints the arable land has declined from about 0.2 ha per capita (1961) to less than 0.15 ha per capita (2003). Due to the GCC countries being in the water scarce region, the major water demand is met through high-cost desalination, which some views as threat to marine environment, as the brine or rejected water affects the marine life through increasing water temperature and increasing water salinity.

The GCC countries in general fall under physical water scarcity (IWMI 2008), that is water resources development is approaching or has exceeded sustainable limits. The GCC countries are situated in the Dryland Systems where mean annual precipitation (P) is less than two thirds of the potential evapotranspiration-PET (evaporation from soil plus transpiration by plants). Hyper-arid areas are considered as true deserts like those in the GCC countries (UAE, Saudi Arabia and Kuwait).

In such sandy desert soils, the lack of water constrains the production of crops, forage, wood, and other ecosystem services.

Due to above climatic constraints, and despite GCC countries being capital rich nations, these countries are facing challenges to combat desertification and for optimum agriculture production. The climate change through rise in temperature, decline in rainfall and increase in evapotranspiration will further impact agriculture. Given these existing and predicted challenges, it is apparent that it would be hard for the GCC countries to achieve food security locally unless there are considerable technological innovations in agriculture research (biotechnology/ molecular biology) to boost production. In the absence of which the food imports will continue for many years ahead. Food import has various aspects of benefits and impact on economies, it gives opportunity to trade with other countries and build up relationships, however, long term food import has some concerns; significant financial obligations – capital flow from food importing countries can affect national economies, no control on food quality – production, high risks of food insecurity during wars and when food import demand increases. Worldwide, almost all countries are importing food to various extents. The current food import by GCC countries is more than 90 %, and self sufficient in some commodities (cucumber, tomato). The UAE imports 85 % of its food (Daniel 2011).

While the food is imported, at the same time the countries are also importing virtual (or embodied) water. This is a measure of the total water used in production of a good or service, e.g., 1 kg of wheat requires about 1,000 l of water, and 15,500 l of water is required to produce 1 kg of edible beef (D'Silva 2011). The concept was initially used to illustrate the advantages to water scarce nations for trade with other nations, rather than attempting to produce all goods locally. The energy-food-water-land complex (GFN 2011) is shown below.

Energy, Food Security, Water and Land Are linked

A solution for one will impact another. Producing more food or energy requires more water. Water options, such as desalination, require more energy. How countries solve this and other puzzles will determine their economic success (GFN 2011).

The GFN (2011) clearly shows the complex between the resources used and generated. It shows 20,000 l of water is required to produce 1 kg of beef, 60 kW h (kWh) are required to produce 20,000 l of desalinated water, and 140 global square meters of land is required for a year to absorb the CO_2 of diesel used to generate 60 kWh of electricity. This can be illustrated in a simple equation:

$$20,000\ liters\ of\ water = 1\ kg\ of\ beef = 60\ kWh\ of\ energy\ consumed$$
$$= 140m^2\ land\ for\ year$$

1.2 Agriculture, Water Stress and Water Resources

Based on the precipitation (P) and potential evapotranspiration (PET) i.e., (P/PET) the dryland systems could be categorized as dry sub-humid (0.5–0.65), semiarid (0.2–0.5), arid (0.05–0.2) and hyper-arid (<0.05) showing an increasing level of aridity or moisture deficit (MEA-Millennium Ecosystem Assessment 2005). The GCC countries are lying under hyper-arid conditions and therefore face severe water constraints for agriculture. The GCC countries face various levels of water stresses, that is Kuwait and UAE are under critical (>10,000 persons per mM^3), Oman slight (<2,500 persons per mM^3), Bahrain and Qatar serious (between 5,000 and 10,000 persons per mM^3) and Saudi Arabia under significant (between 2,500 and 5,000 persons per mM^3) water stresses.

Water is the scarcest resource in the GCC countries. In the water scarce countries agricultural policies are based on the water allocation to agriculture sector. In the GCC countries there is huge pressure on the fresh water resources by the agriculture, and forestry and industries. The GCC countries therefore are reliant on non-conventional water resources (desalinated, treated wastewater, brackish) to offset economic growth and quality of life. Various countries have quantified their water resources and allocations. As per UAE Water Conservation Strategy (MoEW 2010), in the case of UAE total groundwater storage (583 km^3) is distributed into fresh (20 km^3), slightly brackish (190), brackish (148 km^3), slightly saline (225 km^3), whereas water generated from different sources (2.4 km^3) is distributed into desalinated (1.7 km^3), treated wastewater (0.6 km^3), groundwater recharge (0.01 km^3) and groundwater inflow (0.14 km^3).

The UAE's water Municipal and Industrial demand has been increased from 6 mM^3 per year (1968) to 1,464 mM^3 per year (2008). Similarly agriculture area increased from 3,000 ha (1968) to 100,000 ha (2008), including forages on 39 % of farmed area in 2008 and represent 45 % of total agricultural water use (MoEW 2010). The rapid agriculture expansion was abruptly halted in 2000 and slowly declined after, the cultivated area reduced by 4,200 ha per year since 2000. This is mainly due to new policies on subsidies, degradation of groundwater quality, and increased cost of agriculture production culminated into reduced profitability. The area under forest increased from 58,000 ha (1989) to 347,000 ha (2008) and total water use increased from 4 mM^3 (1968) to 694 mM^3 (2007) respectively. In 2008 total UAE water demand was estimated at 4.6 km^3, projected to 7.1 km^3 in 2030 (MoEW 2010). Agriculture being the single largest water user. Agriculture, forestry and urban landscaping use 60 % of UAE's water. Treated wastewater is mainly used in both amenity and landscaping or in a restricted range of agricultural production.

The agriculture in UAE is costing the environment, where, intensive groundwater abstraction during the last 30 years mainly for irrigation purposes in the eastern coast of UAE has caused severe seawater intrusion problems. Consequently, many agricultural farms have either reduced productivity significantly, or they have been abandoned, and only few progressive farmers are using small scale desalination units to desalinate groundwater to allow them continue farming activities. There has been significant impact on agriculture activities in the seawater intrusion vulnerable areas. Under such conditions of high groundwater abstraction, it is essential to monitor the groundwater level to assure that the groundwater is not depleted to a level where combat is not possible. There are various assessment techniques including hydrogeological and geophysical methods, and these methods have been used to determine the current water resources situation and the extent of seawater intrusion in the area. The analysis of current situation indicated that integrated water resources management including new agricultural policy and planning is crucial. In 2010 water conservation strategy (MoEW 2010) for the country which mainly focuses on water demand management has been put into act. The strategy represents a major achievement in the realization of the Government's vision to secure sustainable water resources development for future generations, and will be implemented, monitored and sustained through close coordination with all water sector related partners in the UAE. In the same way Abu Dhabi has prepared Abu Dhabi Water Resources Master Plan (EAD 2009c), and Strategic Plan for Sustainable Wastewater Reuse (EAD 2009d). The UAE Ministry of Environment and Water has recently signed an agreement with Dubai based International Center for Biosaline Agriculture to prepare UAE Agriculture Strategy.

In 2008, a ban on Rhodes grass cultivation in Abu Dhabi emirate was imposed. Farmers Services Center (FSC) – the intermediary between Abu Dhabi Agriculture and Food Safety Authority (ADAFSA) and the Abu Dhabi farming community was created to look after the farmers services in Abu Dhabi emirate. Jointly with the FSC, the ICBA team is demonstrating the value of improved varieties of non-conventional forages (perennials salt-tolerant forages-*Paspalum vaginatum, Distichlis spicata, Sporobolus virginicus and sporobolus arabicus*) for replacing Rhodes grass in the Western region of Abu Dhabi emirate, and through improved and best management farming practices. Rhodes grass is believed to consume high amount of water. Among winter forages, barley, fodder beet, safflower, mustard and quinoa, and summer salt-tolerant crops sorghum, pearl millet guar, cow pea and sesbania have been demonstrated in the farmers fields. The objective of this attempt is to reduce 40 % water use in the Emirate in 5 years (2009–2014). This way huge pressure on water resources will be reduced.

1.3 Ecological Footprint of Consumption and Biocapacity – GCC Countries

In 2007, global humanity's total Ecological Footprint (EF) was 18 billion global ha (gha) with world population at 6.7 billion people (Ewing et al. 2010), the average person's footprint was 2.7 gha, with 11.9 billion gha biocapacity (1.8 gha per person). Thus, it is apparent that humanity uses the equivalent of 1.5 planets to provide the resources we use and absorb our waste. This means it now takes the Earth 1 year and

Table 1.1 Ecological footprint and biocapacity of GCC countries (GFN 2012)

GCC countries*	Ecological footprint per person (gha)	Biocapacity per person (gha)	If everyone in the world consumes like (country in first column) the EF would be (see below) planets
Kingdom of Saudi Arabia	3.99	0.65	2.25
Kuwait	9.72	0.43	5.47
Oman	5.69	2.20	3.20
Qatar	11.68	2.05	6.58
UAE	8.88	0.64	5.00

*Bahrain is one of the GCC countries; however, the data about Ecological footprint and biocapacity is not yet available

6 months to regenerate what we use in a year. This is also reflected from the current global situation, that today more than 80 % of the world's population lives in countries that use more resources than what is renewably available within their own boundaries.

The EF is a measure of how much biologically productive land and water an individual, population or activity requires to produce all the resources it consumes and to absorb the waste it generates, using prevailing technology and resource management practices, it is measured in gha. The gha is normalized to the area-weighted average productivity of biologically productive land and water in a given year. Because different land types have different productivity, a gha of, for example, cropland, would occupy a smaller physical area than the much less biologically productive pasture land, as more pasture would be needed to provide the same biocapacity as one hectare of cropland. The Biocapacity (BC) is the capacity of ecosystems to produce useful biological materials and to absorb waste materials generated by humans, using current management schemes and extraction technologies. The BC of an area is calculated by multiplying the actual physical area by the yield factor and the appropriate equivalence factor, and presented as global hectares.

In the GCC countries, the figures of EF and BC are astonishing (Table 1.1, Figs. 1.1, 1.2, 1.3, 1.4, 1.5, 1.6, 1.7, and 1.8), where the nominal difference between the EF and BC in gha per person is Qatar (9.63), Kuwait 9.29; UAE 8.24, Saudi Arabia 3.34; and Oman 3.49 compared to world figure of 0.92. There is net deficit between ecoresources generated than consumed and wasted. This difference is compensated through food import, and in GCC countries the total bill of food import ballooning from US$8 billion to US$20 billion from 2002 to 2007. From these figures it is predicted that if everyone in the world consumes like (country in first column) the EF would be (see below) different number of earth planets (4th column of Table 1.1; Figs. 1.1 and 1.2).

Looking at the Fig. 1.3 from Kuwait, a sharp decline in gha from just above 6 gha per capita to less than 2 gha per capita just in 1 year may be surprising to those not familiar with what has happened in that year. This sharp decline can be linked to Gulf War when most of the inhabitants (local and expatriate) left the country due to security reasons, therefore, the human demand on the ecosystem was decreased, however, the BC continued to decline. It is worth noticing that gha per capita increased after liberation and peaked in 1998.

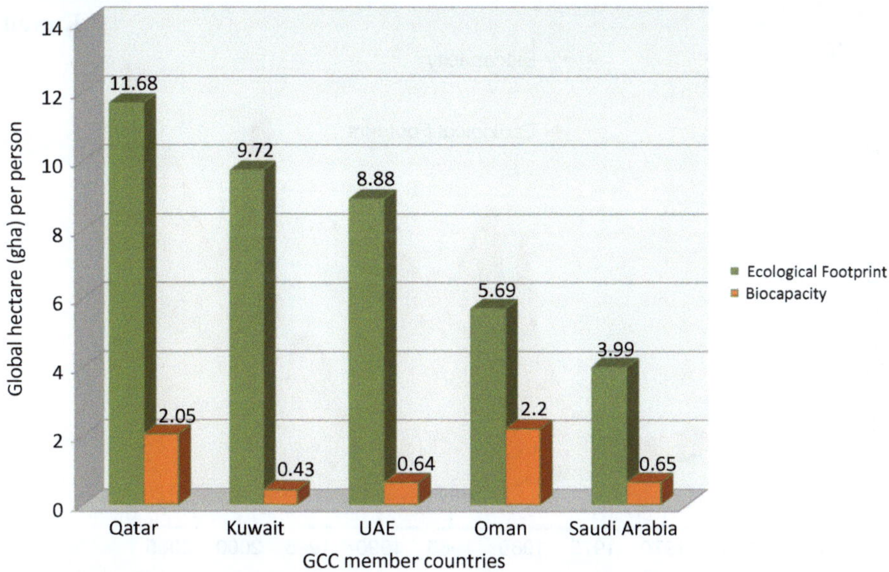

Fig 1.1 Ecological footprint and biocapacity (global hectares per person per year) of GCC countries

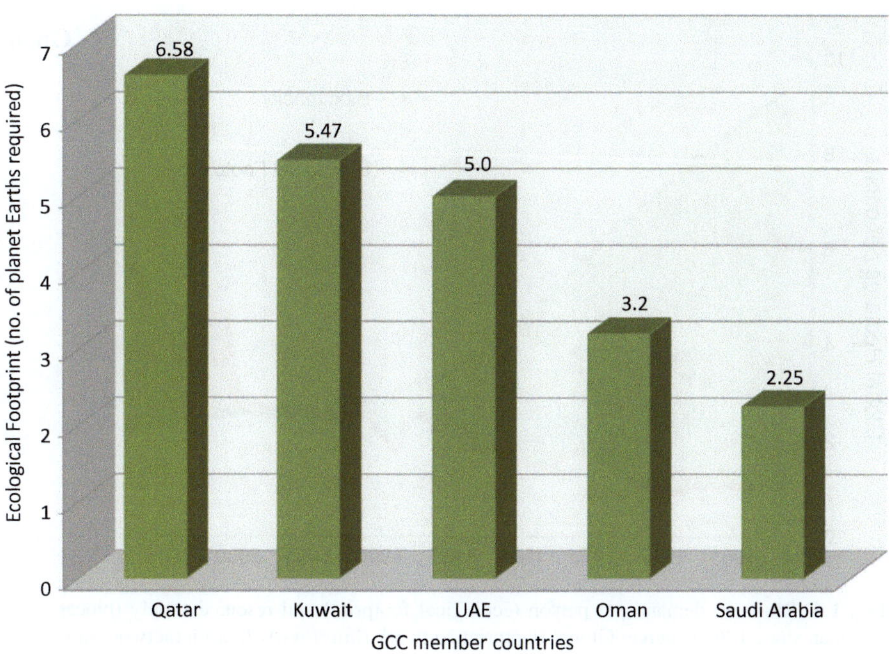

Fig. 1.2 If everyone in the world consumes like each GCC country the EF would be number of earths planets (see above the *bars*), but we have only one

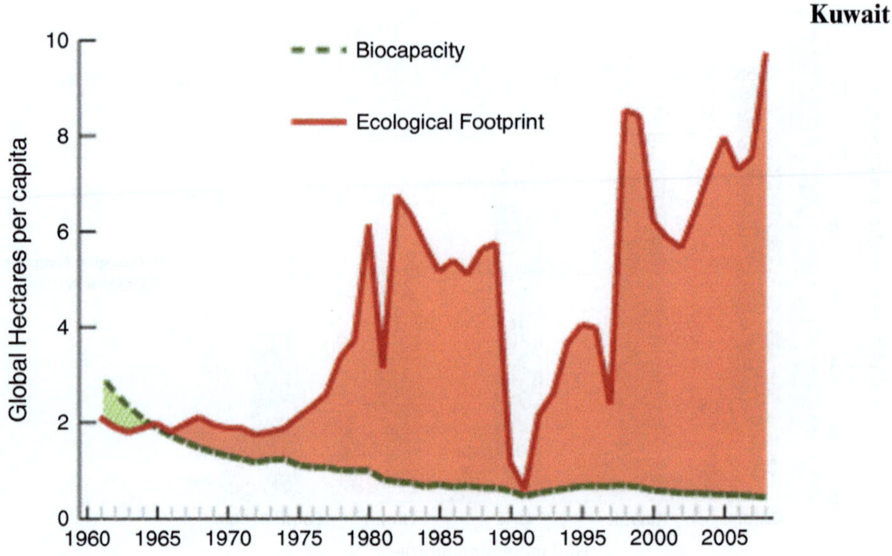

Fig. 1.3 Resource demand per person (ecological footprint) and resource supply (biocapacity) in Kuwait since 1961 (Source: Global Footprint Network (http://www.footprintnetwork.org))

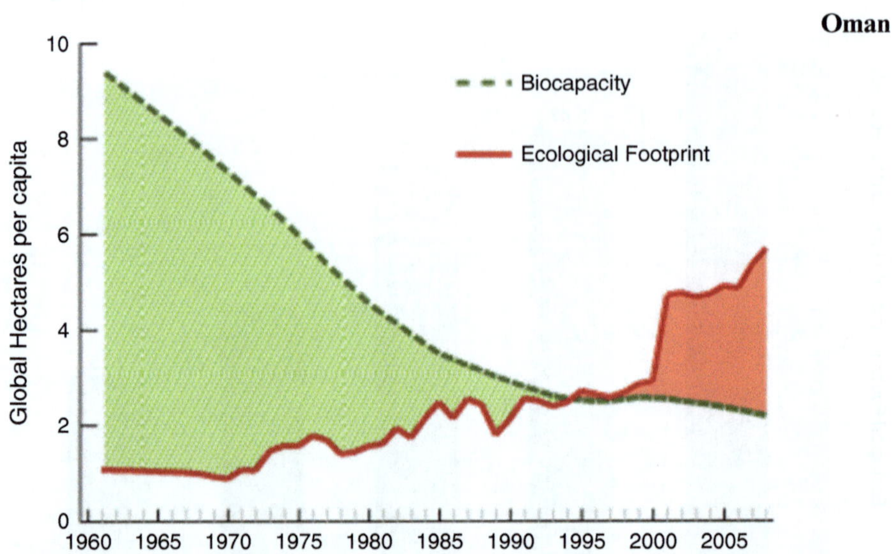

Fig. 1.4 Resource demand per person (ecological footprint) and resource supply (biocapacity) in Oman since 1961 (Source: Global Footprint Network (http://www.footprintnetwork.org))

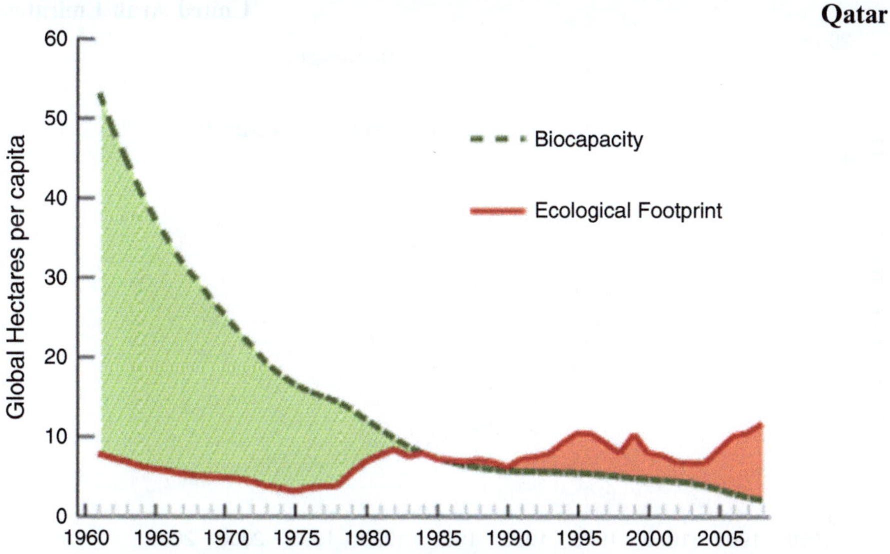

Fig. 1.5 Resource demand per person (ecological footprint) and resource supply (biocapacity) in Qatar since 1961 (Source: Global Footprint Network (http://www.footprintnetwork.org))

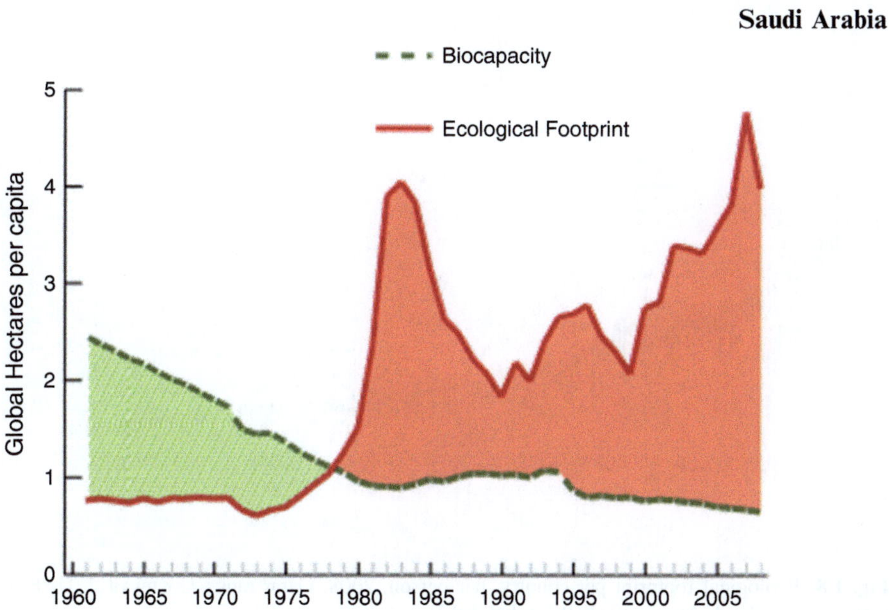

Fig. 1.6 Resource demand per person (ecological footprint) and resource supply (biocapacity) in Saudi Arabia since 1961 (Source: Global Footprint Network (http://www.footprintnetwork.org))

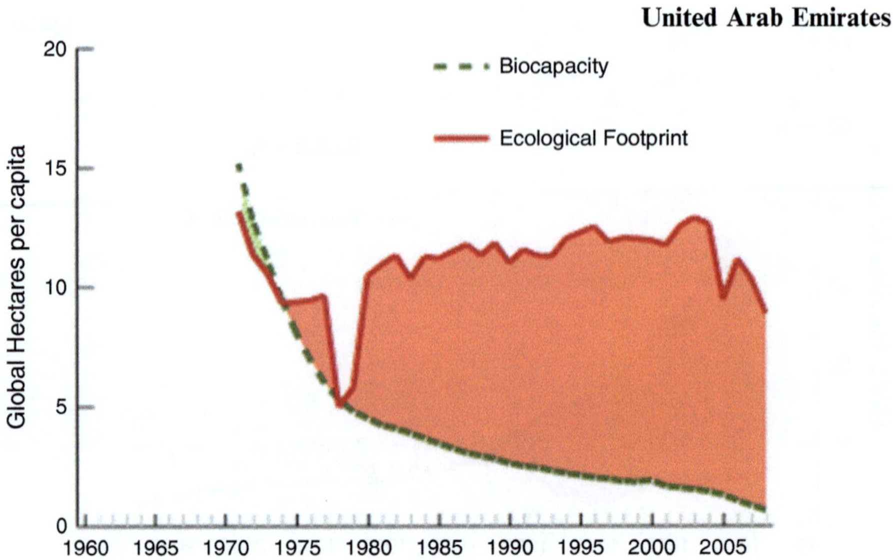

Fig. 1.7 Resource demand per person (ecological footprint) and resource supply (biocapacity) in United Arab Emirates since 1971 (Source: Global Footprint Network (http://www.footprintnetwork.org))

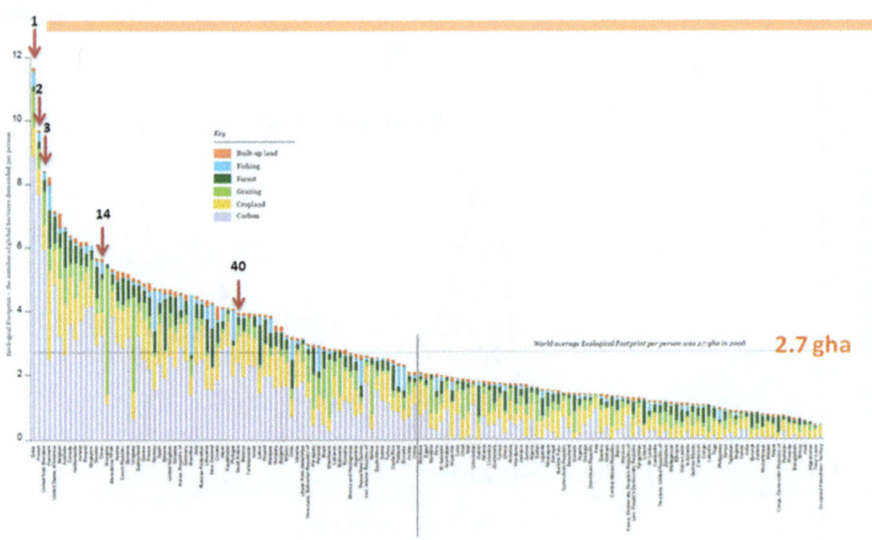

Fig. 1.8 Ecological footprint per country, per person, 2008. Qatar ranks 1, Kuwait 2, UAE 3, Oman 14 and Saudi Arabia 40 (WWF-GFN 2012)

1.4 Prospects of Using Reclaimed Water (Treated Wastewater) in Agriculture

The use of treated wastewater *(called reclaimed water-RW)* for irrigation is a worldwide practice even in water rich countries. Wastewater reuse is conceived as the best discharge option since irrigation requires less treated effluent quality depending on the crops and level of human contact. Moreover, the economic value of applying reclaimed wastewater for agricultural irrigation stems from the nutrient contents that limit application of chemical fertilizers and the amounts of fresh water that could be freed from the agricultural sector. However, to a large extent, this practice depends on the governmental policies and socio-cultural context that varies from country to another and from society to another. The UAE is producing 600 mM^3 of RW by processing municipal and industrial wastewater using tertiary level of treatment. Some 352 mM^3 is used to irrigate landscapes and amenities, while another 248 mM^3 remains unused. The unused water has the potential for groundwater recharge as well as use in industry and agriculture.

With the current population growth rate of the UAE, it is most likely that 1,400 mM^3 of RW will be available by 2030. The quality of RW after tertiary level treatment is generally acceptable as the contaminants causing environmental and human health risks are removed. Exception is for its use to irrigate high value food crops, where the concentration of TDS and *fecal coliforms* are considered higher for unrestricted use. Should the RW is to be opted for growing high value food crops, further purification is required through membrane and ultra filtration technologies, such as those used at the Sulaibiya treatment plant in Kuwait. It is possible, but could be expensive to remove most of the known organic and inorganic (including heavy metals) contaminants beyond the irrigation standards using advanced treatment processes. Alternatively, to reduce the hazards a set of best management practices need to be formulated, including but not limited to adopting remedial or preventive measures related to irrigation technology and management (how, when and how much?); management of cultivated soils; and selection of non-food chain crops.

In Kuwait, the Sulaibiya Wastewater Treatment and Reclamation Plant is considered by far the largest facility of its kind in the world to use reverse osmosis (RO) and ultrafiltration (UF) membrane-based water purification systems. The plant's initial daily capacity is 375,000 m^3, which could be expanded to 600,000 m^3 day^{-1} in the future. It is believed that treated wastewater will contribute to 26 % of Kuwait's overall water demand, reducing the annual demand from non-potable sources from 142 to 26 mM^3 (AFED 2009). Similar to UAE and Kuwait, other countries may have taken the initiatives in such aspects as a way forward to address water scarcity in the respective countries.

1.5 Efforts of GCC Countries for Local Agricultural Production

The GCC countries are continually struggling to increase local agricultural production through various conventional and innovative ways. In parallel they are seeking alternate ways of food security. Successful agriculture and food security based on local agriculture production can only be granted if best soils and sufficient good quality water is available to offset irrigation requirements. In these countries both resources are not sufficient to meet food demand through local agricultural production. It is, therefore, necessary to find ways for efficient utilization of soil resources; this is only possible if soils are characterized by using internationally recognized standards and rationally used based on their potential.

National soil surveys are essential to understand distribution and extent of soils of varying land use capabilities particularly for designating potential agricultural areas. Investigation of soils for a particular purpose involves the ordering of soils into groups (soil classification) having similar characteristics and land use potential. As described by Fitzpatrick (2013), in general, soil-classification systems currently used in most countries involve the use of the following three broad systems (Fitzpatrick 2004); (1) General-purpose broad soil classifications, which communicate soil information at international scales [e.g. Soil Taxonomy (Soil Survey Staff 1999; 2010) and World Reference Base (2006)] and national scales [Australian (Isbell 1996)]; (2) State, provincial or regional soil classifications, which are designed both to assist with "user-friendly" communication of soil information and to account for the occurrence of soils that impact on existing and future industry development and prosperity [e.g. Western Australia (Schoknecht 2001) and South Australia (Hall et al. 2009)]; (3) Special-purpose and more technical classification systems, which are used for local or single purpose applications. These systems generally involve using detailed soil-assessment criteria with recommendations for specific soil-management practices for a range of specific industries [e.g. use of Acid Sulfate Soils in mineral exploration (Skwarnecki and Fitzpatrick 2008)], land evaluation for irrigated agriculture purpose (FAO 1976), and to key out of soils of United Arab Emirates (Shahid et al. 2013a; 2014).

Acknowledging the needs of national soil inventories for informed decisions to aid land use planning, and seeking areas having highest potential for irrigated agriculture, almost all GCC countries have completed their national soil inventories. The Kuwait (KISR 1999a, b; Shahid and Omar 1999; Omar and Shahid 2013), Oman (Ministry of Agriculture and Fisheries 1990), Qatar (Ministry of Municipal Affairs and Agriculture 2005; Scheibert et al. 2005), Saudi Arabia (Ministry of Agriculture and Water 1985), and UAE (EAD 2009a, b; 2012; Shahid et al. 2013b) have completed their national soil inventories (published soil and other thematic maps for land use planning). The most important of these is the land suitability maps for future irrigated agriculture expansion to meet food demand of growing population. They identified areas (Saudi Arabia 13.7 %); Oman (7.07 %), Kuwait (35 %), Abu Dhabi emirate (5.4 %) and

Qatar (3.9 %) as highly to moderately suitable for large-scale irrigation farming. The question is do the GCC countries have sufficient water resources to farm these areas?

The soil database and soil information systems (KISR 1999a, b; EAD 2009a, b) established during these soil inventories is crucial for speedy development of Kuwait and UAE. Therefore, it is essential that the soil surveys data must be integrated into the Government information system for informed decisions and proper land use planning. The pioneering efforts of Kuwait in developing soil information system (Soil Information System) and UAE (UAE Soil Information System) are highly acknowledged.

Oman has furthered investigation on the management of salt-affected soils and water for sustainable agriculture (Al-Rawahy et al. 2010). Later, in an effort to understand the real picture of damage and to rethink of salinity management in Oman, the Ministry of Agriculture and Fisheries jointly with other organizations in Oman and with technical support from ICBA has prepared Oman Salinity Strategy "National Strategy to Combat Salinity and Protection of Water Resources from Pollution and Salinity in the Sultanate of Oman". The Oman Salinity Strategy-OSS (MAF 2012) aimed to identify ways to control water resource deterioration due to salinity and pollution and to develop sustainable, economical methods to utilize resources for optimum agriculture production.

1.6 Potential Options of Food Security in GCC Countries

There is limited scope for expansion of arable land in the GCC countries, and the emerging threat to existing agriculture from climate change in the form of unpredictable weather, floods, and other disastrous events makes the task of providing enough food for the existing and growing population even more challenging. Given these existing and predicted challenges the GCC countries are seeking new potential options for food security. There are many ways by which these countries can achieve food security, like intensification in local food production using high-tech multipronged approach (rational use of soil resources, modern irrigation systems, protected agriculture, water conservation, sector wide water reforms, use of alternative water sources); continuing food import; outsourcing food production to countries which have comparative advantage for agricultural expansion; leasing farmland abroad, and through creation of ***GCC Countries Strategic-FOOD RESERVE*** to be used in case of emergency for at least 2–3 months. There are however, strengths, weaknesses, opportunities and threats (SWOT) of achieving food security by above ways, which should be properly examined prior to final decision.

One of the potential options to secure food is through leasing land in other countries. Driving forces behind land lease are: (1) to secure food supply; (2) the surging demand for agro-fuels and other alternative energy sources and; (3) the sharp rise in investment in both the land market and the soft commodities market.

Many view such deals as threats to food security of host nations, others, see as opportunity to help poor nations through providing jobs, training in modern agriculture technologies, and developing infrastructures. This trend has come under heavy scrutiny since mid-2008. Investment in agricultural land is thought to mean to provide food to needy nations. Many claim this approach conflicts with the urgency of increasing domestic food supplies in the world's poorest and most vulnerable countries. Some view, no matter how convincing the claim that the global land deals will bring much-needed agricultural investment to poor countries, evidence shows there is simply no place for the small farmers. According to Daniel (2011), a dangerous element of the land grab trend is the shift from domestic to foreign control over food resources and food-producing lands. Others view, land deals diminish the possibility of reaching food self-sufficiency for poor nations and some view land concessions as governments out-sourcing food at the expense of their most food-insecure citizens, importantly, most of the target or "host" countries themselves are net food importers or even emergency food aid recipients. Views of some policy makers are shared below.

The FAO Chief expressed his concern about the potential consequences of swift land grabbing on political stability, has said he supports the proposed Gulf food deals as a means of economic development for poor countries. If the deals are constructed properly, he said, they have the potential to transform developing economies by providing jobs both in agriculture and other supporting industries like transportation and warehousing (Coker 2008). The IFAD's President expressed hope for possible development opportunities through land purchases. When such deals take into account interests of both parties they help increase agricultural production in developing countries, provide jobs, boost export and bring in new technologies to improve farm efficiency (Kovalyova 2008). IFPRI calls for a code of conduct both for foreign investors and the host countries in order to protect the interests of small farmers, as well as address environmental concerns on biodiversity and water and land resources stemming from the impact of large-scale farmland investments (von Braun and Meinzen-Dick 2009).

Regardless of positives and negatives of land deals, many deals have been reported (von Braun and Meinzen-Dick 2009), including the Gulf deals for farmland abroad:

Saudi Arabia (Sudan-wheat, vegetables, animal feed; Indonesia-rice; Egypt-barley, wheat, livestock feed; World-agriculture projects)
UAE (Sudan-corn, alfalfa, wheat, potato, beans; Pakistan-agriculture; Ethiopia-tea)
Qatar (Kenya-fruits, vegetables; World-food and energy; Vietnam-agriculture)
Kuwait (Cambodia-rice)
Bahrain (Philippine-agroforestry; Turkey-agriculture); Dubai World Trading Company (Ethiopia-tea).

Leasing prime land could be a threat to food security of host nations. Therefore, to have *win-win* scenario between investors and the host nations, a compromising scenario has to be developed, which is respected, supported and owned by everyone in the land deals chain. We propose the compromised scenario as below:

Win-win Perception of Land Deals

Worldwide over one billion hectares are of marginal quality (salinity & sodicity) lands set aside and currently not in use, these exist mostly in poor developing countries (Shahid 2013a), or globally 10 % of the total arable land is affected by salinity and sodicity, extending over more than 100 countries (Kovda and Szabolcs 1979). The full exploitation of marginal lands requires an integrated approach of reclamation (physical, chemical, hydrological, and biological techniques) to understand the processes, establish the cause-effect relationship, and develop appropriate methods of constraint/stress alleviation, soil restoration and quality enhancement to increase crop production in marginal lands (Shahid and Rahman 2011). The marginal lands are mostly distributed in poor developing countries, financially not sound and therefore cannot afford to bring these lands into agriculture production system. This is the niche for foreign investors to lease such lands for agriculture production. Based on the level of degradation, such soils may not be used straight away, but through using reclamation techniques, and biosaline agriculture (growing salt-tolerant crops, such as forages), these can be brought into production (Shahid 2011, 2013b). This scenario will leave the prime lands for the host nations to live on.

If this scenario is well received and have wider acceptance from all stakeholders. Then we have to look scientific institutes who are specialised in dealing with marginal lands. In this respect the Dubai based International Center for Biosaline Agriculture (ICBA) is unique in the world by being non-profit, non-commercial, international organization and deals full time applied biosaline agriculture aspects to utilize marginal saline land and water resources. The ICBA uses integrated approach of soil reclamation to improve degraded lands for sustainable agriculture production. ICBA network is in all Islamic countries who are the members of the Islamic Development Bank.

1.7 Inception of ICBA

In 1999 the International Center for Biosaline Agriculture (ICBA) was established in Dubai to conduct biosaline agriculture research and implement in Islamic Development Bank (IDB) member-countries including the Gulf States (GS). Since ICBA inception in 1999 and until 2009, ICBA mission was to demonstrate the value of saline water resources for the production of environmentally and economically useful plants and to transfer the results to national research services and communities regionally and globally. In the second strategic plan (2009–2014) ICBA is focusing on helping water-scarce countries improve productivity, social equity and environmental sustainability of water use through an integrated water resource systems approach, with special emphasis on saline and marginal quality water. ICBA is unique in the world as it is involved full time in Biosaline Agriculture activities. Further information about ICBA activities, network and achievements can be seen in Shahid et al. (2011).

1.8 Soil Use and Protection Strategies in the GCC Countries

Soil is a vital part of our environment and has a key role in environmental interaction, linking the atmosphere, water resources and land use. There is a strong need to shift our thinking about soil more toward the environment and ecosystems and water management, away to a certain extent from the agronomic paradigm. This requires contribution of soil to multi-discipline studies, communication with policy makers and decision makers, to enhance the soil image among a public concerned with resource uses and a need of broader future concepts of soil by applying modern technologies. Soil is probably the most precious non-renewable resource on the planet Earth, its destruction (mining, deforestation, unsustainable farming, and pollution) poses more serious, immediate and irreversible problems to humanity in terms of food production, water supplies and green house gasses than depletion of fossil fuels. To save the precious resource and for long term sustainable services, the ministries and government departments should incorporate soil use and protection strategies into their business plans regarding agriculture, forestry, water resources management, waste disposal and land use planning. They must stress on the rational use of soil resources based on their potential for a specific land use.

The need to protect soils against pollution has been realized by many countries, which led the government to set policies for soil protection. Although the need for soil protection has been realized much later than for clean air and drinking water, it has come to force through horrific cases which have shown the soil is not capable of absorbing and detoxifying unlimited amounts of waste materials from agriculture, industry and society at large. The pollutant commonly arises from the disposal of industrial waste products in designated landfills or uncontrolled dumps, from spillage and atmospheric fallout, industrial release of contaminated waters, etc. Even the use of fertilizers, pesticides and non-conventional waters in agriculture increase amounts of certain contaminants. Protecting the activity of the soil to support agriculture is an important part of wider efforts to promote the economies of farmers, and social progress for people. A number of national and international organizations have recently included soils in their strategies as a natural resource of vital importance, be it for agricultural production, water management, and a resource for many uses. Soils have multiple functions that are vital to global sustainability of the earth as a living system and basis for human survival.

The way we use our soils, and the influence of our activities on it, will have important and far-reaching impact on the quality of soil environment. The GCC countries being in the dryland systems are highly vulnerable to land degradation, and therefore, there is an urgent need to develop national soil protection strategies to assure they provide services on a sustainable basis and for a long time. This will lead to the best overall soil services for various purposes including crop production, and inevitably include many different social, economic and environmental objectives. Sandy desert soils or wrongly selected soils for irrigated agriculture require heavy inputs due to very low water holding capacity, high drainage and low inherent soil fertility. These soils under hot climate conditions require irrigation on a daily basis causing nitrate (NO_3) pollution of groundwater. In many cases water is not available at site but transported from desalinated plants which add cost further in addition to pressure on drainage and contamination of groundwater. There is a need to assess soil resources through expert land evaluation system to identify areas having the highest potential for irrigated agriculture and maps prepared. This map is then to be overlaid on available water map to find area where both good quality soil and sufficient water is available for agriculture. This way the scarce water resources can be optimized for a productive and profitable agriculture. This is only achievable when sufficient water and soil data of the area of interest is available and the soil and water database is integrated to facilitate decision makers.

1.9 Environment Impact Assessment (EIA) and Development of Action Plan

Many irrigated agriculture based projects failed due to the absence of EIA studies conducted prior to the project implementation. The EIA is a tool for sound decision-making; it is a formal process for ensuring that potential environmental impacts are considered in major project approval. This does not imply that decision will be made solely on the basis of minimizing adverse environmental impacts. Decision makers will also consider the economics and social concerns, as well as the political acceptability of the project. What the EIA process does is that it allows environmental concerns to stand at the forefront. The EIA also ensures that the project proposal is designed to mitigate adverse environmental consequences as far as possible. Thus, as a result of the EIA process, developments, which go ahead, should be designed to be environmentally friendly. The Pre-EIA indicates possible impacts of the project on the environment and help design the project in an environment friendly manner. The Post-EIA depicts adverse impacts of the project on the environment. The post EIA will allow determining soil characteristics and properties, contamination with inorganic constituents, and changes in biodiversity, after the project has been completed.

1.10 Agriculture and Climate Change, Implications on the Environment in the GCC Countries

Properly managed agriculture sequesters carbon and mitigates impact of climate change. However, agriculture also contributes to climate change in the form of GHG emission through fertilizers use (release of nitrous oxides), enteric fermentation (methane-CH_4 emission) and rice cultivation (CH_4 methane emission). The global warming potential (Scherr 2011) of different GHGs is different (e.g. CO_2 1; Methane-CH_4 25; Nitrous oxide-N_2O 298, and Hydro fluorocarbons-HFCs from 124 to 14,800). The percent share (IPCC 2007; Scherr 2011) of different sectors to GHGs is Agriculture 13.5, Buildings 7.9, Energy 25.9, Industry 19.4, Forestry 17.4, Transport 13.1 and Wastes 2.8.

It is therefore, realistic to state that agriculture is both problem and solution for climate change. At least 13.5 % of global GHG emissions come directly from the farm sector. That is more than transport and not far behind the contribution from industry; therefore, coherent and effective policies are needed to mitigate the impact of climate change. Devising policies and undertaking commitments to mitigate climate change through agriculture clearly means knowing more about agriculture's carbon footprint and environment cost of producing food in water scarce region. With the intention to intensify agriculture in the GCC countries, it is likely that the carbon footprint will increase, since farming is set to expand to produce more food for a growing population. Exact estimates in the region to increase food production do not exist, however, it is projected that world food production will need to double from current levels if projections of more than nine billion people in 2050 prove correct. That means more land-use change, more cultivation, more livestock and more use of fossil fuels and hence more carbon footprint.

The Regional Circulation Models (RCM) projection for precipitation change in the GCC countries predicts the region to get drier, with significant rainfall declines in the wet season outweighing slight increases during the drier summer months (Cruz et al. 2007; Hemming et al. 2007). Meanwhile the rainfall will become more unpredictable with the region experiencing an increase in extreme rainfall events (Hemming et al. 2007). Such events have been witnessed in Kuwait in 1998 (70 mm rain in an hour) and the Kuwait government declared emergency, in Jeddah Saudi Arabia in 2009, rainfall inundated large urban areas with some death tolls. The 2007 hurricane Gonu in Oman has left its footprint in the coast as well as in Muscat city of Oman. Let us assume that rainfall does not decrease, which is less likely, the rise in temperature will still have significant impact on evapotranspiration and water balance, reducing available resources up to 15 %, whereas the increase in agricultural water demand by the year 2020 is estimated at 6 % (Bou-Zeid and El-Fadel 2002).

A simulation carried out by Boston University revealed that 1 m Sea Level Rise (SLR) would directly impact 41,500 km^2 of the Arab Coastal land (Egypt, Tunisia, Morocco, Algeria, Kuwait, Qatar, Bahrain, and the UAE). Almost 11 % of the land area of Bahrain will be lost due to 50 cm SLR if no action is taken for protection (Al Janeid et al. 2008). The land mass of Qatar, UAE, Kuwait will be most vulnerable, in these countries under different projections and scenarios 3 % land will be affected (1 m SLR), 8 % (3 m SLR) and 13 % (5 m SLR). Qatar is by far the most exposed country. The GDP of Qatar and UAE is at risk as over 2 % of their respective GDPs are at risk for 1 m SLR, increasing to between 3 and 5 % for 3 m SLR. The low-lying coastline exists in most of the GCC countries (UAE, Kuwait, Qatar, Oman, Bahrain) which may be within the reach of few meters sea level rise, e.g., in the UAE which has nearly 1,300 km of coastline, about 85 % population and 90 % of infrastructure are located within several meters of sea level (Fencl et al. 2009). In the UAE most of the agricultural activities are in the inland, therefore, it is less likely that SLR will affect these farms, but the rising temperature and decrease in rainfall will. Two scenarios (Fencl et al. 2009) of SLR were used on the Abu Dhabi Coastal Ecosystem base map; (1) no accelerated ice cap melting and by 2050 1 m above mean sea level (MSL); (2) accelerated ice cap melting, 3 m above MSL by 2050 and 9 m above MSL by 2100. Much of Abu Dhabi is above 1 m, therefore most of the area is not vulnerable to 1 m rise, but low lying areas; under scenario 2, with 3 m SLR several offshore islands, mangrove village and industrial city south of main land will be under water; with 9 m of sea level rise Abu Dhabi will look fundamentally different as a city and society as of today. The extent of inundation in Abu Dhabi Emirate (Fencl et al. 2009) range from 344 km^2 (1 m SLR), 804 km^2 (3 m SLR) and 1,672 km^2 (9 m SLR), whereas in Dubai it is 14 km^2, 147 km^2 and 221 km^2 inundation area with 1, 3 and 9 m SLR scenarios respectively. The derived Digital Elevation Model (DEM) from topographic maps shows 332 km^2 areas below 10 m in three emirates (Sharjah, Ajman, Umm Al-Quwain) are highly vulnerable to SLR. It is projected that 1 m SLR would inundate 8.1 % (Ajman), 1.2 % (Sharjah) and 5.9 % (Umm Al-Quwain) area, with 8 m SLR the inundated area would be 24 %, 3.2 % and 10 % respectively (Tolba and Saab 2009).

1.11 Mitigation and Adaptation

Mitigation includes strategies to reduce GHGs sources and emissions and enhancing greenhouse gas sinks. The Adaptation is an adjustment in natural or human systems in response to actual or expected climatic stimuli or their effects, which moderates harm or exploits beneficial opportunities. One of the answers is carbon sequestration, as soil literally captures and absorbs carbon and so offsets emissions not only from farming, but from other sectors too. As for mitigation, policies aimed at reducing emissions in agriculture may be more cost-competitive than in some industrial and transport options.

The GHG emissions from human activity will have to decrease globally from 1990 levels by at least 50 % by 2050 if future global warming is to be limited to a 2 °C temperature increase, as currently recommended by the Intergovernmental

Panel on Climate Change (IPCC). This imperative was repeated by world leaders at the UN Climate Change Conference in Copenhagen in December 2009. Improved analysis of adaptation technologies is required to show how they can contribute to building adaptive capacity and resilience in the agriculture sector.

With the right technologies and systems, improved cropland and grazing land management, restoration of degraded lands and land-use change, such as agro-forestry, can make a major contribution to limiting greenhouse gases. Emissions from livestock production can be reduced by improving nutrition and manure management. Science offers promising solutions in areas such as genetics, so-called second generation biofuels that compete less with land used for food crops, and carbon capture, though clearly more research is needed. Genetics, for instance, could help reduce methane from animals too. Cattle and sheep's sizeable share of greenhouse gases–indeed, methane is many times more potent than carbon dioxide as a greenhouse gas–can offset to some extent by capturing the carbon in pastureland. Better farming methods can help mitigate climate change while also improving water quality, biodiversity and soil quality. It may mean policies to discourage certain kinds of heavily emitting agriculture to reduce risks and slow down land-use changes. In any case, such policies will also relieve other environmental stresses, including in water supply, and make agriculture more sustainable. Incentives for more efficient water use and compensation for vulnerable groups: such measures, if taken together and adapted to specific country situations, would create coherent policy packages to limit agriculture's contribution to climate change, improve the environment and boost the effectiveness and value-added of the farming sector. There is an urgent need for improved climate modeling and forecasting that can provide a basis for informed decision-making and the implementation of adaptation strategies. This is vital for building adaptive capacity and decision-making processes. This information needs to be compiled and disseminated for a range of stakeholders from local to national levels. Relationships between policy makers, researchers and communities should be built so that technologies and planning processes are developed in partnership, responding to producers' needs and integrating their knowledge.

1.12 Improving Extension Services Using Information Technology (*e*strategy)

The Ministries of Agriculture and Fisheries and Agricultural Departments in the GCC countries are responsible for providing extension services to the farming community. The farmers awareness of modern methods and new technologies is achieved through site visits of extension staff, distribution of pamphlets, brochures, meetings, field days and TV talks etc.. Through these means the farmers gain general information, however, the farmers lack farm based technology package to address agricultural related issues on his particular farm. This requires selection of representative farm sites in the agricultural zones where these technologies can be demonstrated to larger farming community. Prior to demonstration, it is essential to conduct an intensive farm resource inventory (soil, water, crop history, machinery, socio-economic, marketing aspects, infrastructures etc.) of the selected farms and the information recorded. Based on this information and the market demand a package of technology (selection of crops and sowing techniques, fertilizer, water and amendment requirements, soil conservation and other necessary management) be developed and demonstrated at pilot scales. The extension staff then works closely with the farmers and adds latest information on designated Ministry website (**e**strategy), where farmer should be given access to the latest updates on agricultural technologies. In addition a mobile service could be an option for fast technology transfer and to give advisory services to the farmer on toll free mobile service. The ultimate objective is to enable farmer to effectively use farm land according to their potentialities, and without endangering their capabilities for production in future. Later **e**strategy can be extended to a larger farming community. This way agricultural enhancement in the GCC countries can be made possible.

1.13 Greater Regional Co-operation and Creation of a Regional Food Reserve (Especially for Staples) to Manage Risks and Uncertainties in Commodity Markets

The above facts and discussion clearly illustrate that the GCC countries are located in water scarce region, hyper-arid climate, insufficient water and arable lands that is leading the region for heavy food import and financial obligations. Continuous food import can only be assured if sufficient capital is available, and the agreements are made with countries from where food is being imported. Should the link between countries weakened, or the countries from where food is being imported go through climatic disasters (floods, storms, drought etc.) or political instabilities, famine and the food available in those countries is not even enough to feed the local population, or not allowed to export, the surety of food import thus become questionable. To avoid such emergencies and to secure the food for the GCC countries, it is visualized to establish "*Strategic Gulf Food Reserve*" and food stocked to be used for at least 2–3 months through mutual agreement within GCC countries.

1.14 Establishment of Gulf Salinity Forum (GSF)

Salinity either of water or soil is threatening agriculture in the GCC countries. The underlying philosophy of the GSF is that salinity issue in the region will not be dealt with adequately unless we think beyond the level that created the problem. So far it has been found that salinity problem is increasing with time, and no one country is devoid of salinity problem. The water scarcity, hyper-arid climatic conditions and salinity issues are collectively threatening agriculture development and environment in the GCC countries, and therefore the formation of GSF become the focus to collectively think about the problem to manage it on a sustainable basis without compromising the quality of the environment and natural resources. In the GSF we will show, how close salinity is linked to agriculture farms, groundwater resources, ecosystems, the environment and human being, and present new opportunities for sustainable soil and water resources uses on an eco-friendly basis. The forum intention is to draw together the people, resources and tools used to solve salinity issues at the regional level. The forum will help facilitate information processing and networking in the Gulf region. Major activities of the GSF will include meetings to improve coordination between forum members, organize awareness days, brainstorming, workshops, exchange of information through organizing conferences, seminars, symposia, workshops and trainings. Publish quarterly eNewsletter, and biannual Gulf Journal of Salinity Management. The forum will have official website which will be regularly updated. The Ministries, Universities, Research Institutes, Environment Agencies and Government Departments etc. from the GCC countries can join the GSF as an institute. The Researchers, Scientists, University Academic Staff, students, etc. can join as individual member. GSF is suggested to be governed by the advisory board. International – 2; GCC Countries – 6 (One from each country).

1.15 Conclusions

The GCC countries are water scarce, have hot environment and vulnerable to climate change. Owing to these conditions the difference between EF and BC is increasing since 1960. These countries are therefore importing major portion of food and have significant financial commitments. Open field agriculture is only possible through heavy environmental cost, and therefore, there is potential of protected agriculture in the region. Whereby the region can be self sufficient in certain commodities (tomato, cucumber etc.), however, production of cereals will be a dream to become true through innovations in agricultural technologies (heat, drought and salt tolerant varieties), waterless agriculture and through climate smart agriculture. Other options are outsourcing food production in countries that have sufficient good quality soil and water resources, and through creating a strategic GCC countries food reserve. Other realistic option is to lease marginal saline lands which are set aside since many years and are not currently in use. These saline lands can be brought under production through using an integrated soil reclamation approach, whereby, ICBA can help to achieve this target. This is a win-win scenario when the prime land is left for the host countries to use for food production.

Acknowledgements We would like to thank and acknowledge Faith Flanigan (World Outreach Coordinator) of Global Footprint Network to supply and granting the permission to use graphs of the Kuwait, Oman, Qatar, Saudi Arabia and UAE in this publication. The ICBA management deserves special thank for keen interest in Gulf Research Meeting 2012 and approving the mission of Dr Shabbir Shahid (Lead Workshop Director) to conduct Agriculture workshop in Cambridge University.

References

AFED (2009) Arab environment: climate change – impact of climate change on Arab countries. Mostafa KT, Najib WS (eds) Report of the Arab Forum for environment and development. Technical Publications and Environment & Development Magazine, Beirut

Al Janeid S, Bahanacy M, Basr S, El Raey M (2008) Vulnerability assessment of the impact of sea level rise on the Kingdom of Bahrain. Mitig Adapt Strat Glob Change 13(1):87–104

Al-Rawahy SA, Ahmed M, Hussain N (2010) Management of salt-affected soils and water for sustainable agriculture. In: Ahmed M, Al-Rawahy SA, Hussain N (eds) Monograph on management of salt-affected soils and water for sustainable agriculture. Sultan Qaboos University, Muscat, pp 1–8

Bou-Zeid E, El-Fadel M (2002) Climate change and water resources in Lebanon and the Middle East. J Water Res Plan Manag 128(5):343–355

Coker M (2008) UN Food Chief warns on buying farms. The Wall Street Journal. 10 September 2008. http://online.wsj.com/news/articles/SB122098802643115897

Cruz RV, Harasawa H, Lal M, Wu S (2007) Chapter 10: Asia'. In: Parry ML, Canziani OF, Palutikof JP, van der Linden PJ, Hanson CE (eds) Climate change 2007: impacts, adaptation and vulnerability. Contribution of Working Group II to the fourth assessment report of the Intergovernmental Panel on Climate Change. Cambridge University Press, Cambridge

D'Silva J (2011) Food price rises and the meat connection. Newsletter of the World Forum on Climate Change Agriculture and Food Security (WFCCAFS) 1(2):2

Daniel S (2011) Chapter 2: Land-grabbing and potential implications for world food security. In: Behnassi M, Shahid SA, Silva JD (eds) Sustainable agricultural development. Springer, Dordrecht/Heidelberg/New York/London, pp 25–42

EAD (2009a) Soil survey of Abu Dhabi Emirate. Extensive survey. Environment Agency-Abu Dhabi, Abu Dhabi

EAD (2009b) Soil survey of Abu Dhabi Emirate. Intensive survey. Environment Agency-Abu Dhabi, Abu Dhabi

EAD (2009c) Abu Dhabi water resources master plan. Environment Agency-Abu Dhabi, Abu Dhabi, 220 pp

EAD (2009d) Strategic plan for sustainable wastewater reuse. Environment Agency-Abu Dhabi, Abu Dhabi

EAD (2012) Soil survey of Northern Emirates. Environment Agency, Abu Dhabi. 3 volumes

Ewing B, Moore D, Goldfinger S, Oursler A, Reed A, Wackernagel M (2010) The ecological footprint Atlas 2010. Global Footprint Network, Oakland. http://www.footprintnetwork.org

FAO (1976) A framework for land evaluation, FAO soils bulletin no. 32. Food and Agriculture Organization, Rome

Fencl A, Elasha BO, Swartz C, Yates D, Fisher J, Klein R (2009) Climate change – impacts, vulnerability and adaptation. Environment Agency – Abu Dhabi, Abu Dhabi, 197 pp

Fitzpatrick RW (2004) Classification systems: Australian. In: Daniel Hillel (Editor-In-Chief) Encyclopedia of soils in the environment. Elsevier, Oxford, pp 211–216. © 2005. ISBN (Set): 0-12-348530-4

Fitzpatrick RW (2013) Chapter 2: Demands on soil classification and soil survey strategies: special purpose soil classification systems for local practical use. In: Shahid SA, Taha FK, Abdelfattah MA (eds) Developments in soil classification, land use planning and policy implication- innovative thinking of soil inventory for land use planning and management of land resources. Springer, Dordrecht/Heidelberg/New York/London, pp 51–83

Global Footprint Network (GFN) (2011) What happens when infinite-growth economy runs into a finite planet. Global Footprint Network 2011 annual report. Oakland, California 94607-3510 USA

Global Footprint Network (GFN) (2012) Ecological footprint Atlas. http://www.footprintnetwork.org/en/index.php/GFN/page/footprint_for_nations/

Hall JAS, Maschmedt DJ, Billing NB (2009) The soils of southern South Australia. The South Australian land and soil book series; vol 1. Geological Survey of South Australia; Bulletin 56, vol 1. Department of water, land and biodiversity conservation, Government of South Australia

Hemming D, Betts R, Ryall D (2007) Environmental stress from detailed climatic model simulations for the Middle East and Gulf Region (ref. MOEN/04/04/02/02b) completed by the Met Office Hadley Center for UK Ministry of defense project 'defense and security implications of climate change'

IPCC (2007) Climate change 2007: impacts, adaptation and vulnerability. Working group II contribution to the fourth assessment report of the Intergovernmental Panel on Climate Change – summary for policymakers and technical summary. WMO-UNEP publication

Isbell RF (1996) The Australian soil classification. CSIRO, Melbourne

IWMI (2008) Areas of physical and economic water scarcity. UNEP/GRID-Arendal Maps and Geographic Library

KISR (1999a) Soil survey for the State of Kuwait – vol II: reconnaissance survey. AACM International, Adelaide

KISR (1999b) Soil survey for the State of Kuwait – vol IV: semi-detailed survey. AACM International, Adelaide

Kovalyova S (2008) UN agencies see "win-win" farmland deals. Reuter, April 19

Kovda VA, Szabolcs I (1979) Modelling of soil salinization and alkalization. Agroké miaés Talajtan 28(Suppl):1–208, Budapest

MAF (2012) Oman salinity strategy. Ministry of Agriculture and Fisheries, Muscat

Millennium Ecosystem Assessment (MEA) (2005) Ecosystems and human well-being: current state and trends. Island Press, Washington, DC.

Ministry of Agriculture and Fisheries (MAF) (1990) General soil map of the Sultanate of Oman. Ministry of Agriculture and Fisheries, Muscat

Ministry of Agriculture and Water (MAW) (1985) General soil map of the Kingdom of Saudi Arabia. Ministry of Agriculture and Water, Riyadh

Ministry of Municipal Affairs and Agriculture (MMAA) (2005) Soil classification and land use specifications for the State of Qatar. Phase one report, vol I: soil survey report; vol II: technical data. Doha, State of Qatar

MoEW (2010) United Arab Emirates water conservation strategy. Ministry of Environment and Water, Abu Dhabi

Omar SAS, Shahid SA (2013) Reconnaissance soil survey for the State of Kuwait. Chapter 3 In: Shahid SA, Taha FK, Abdelfattah MA (eds) Developments in soil classification, land use planning and policy implications - innovative thinking of soil inventory for land use planning and management of land resources. Springer, Dordrecht/Heidelberg/New York/London, pp 85–107

Scheibert C, Stietiya MH, Sommar J, Abdalla OES, Schramm H, Al Memah M (2005) The atlas of soils for the State of Qatar. Ministry of Municipal Affairs and Agriculture, Doha

Scherr SJ (2011) Agriculture – land use and climate change. Farming Outlook 10(1):4–9

Schoknecht NR (ed) (2001) Soil groups of Western Australia. Resource management technical report 193. Agriculture Western Australia, South Perth

Shahid SA (2011) Food security – farmland investment in developing countries, an issue or an opportunity – win-win scenario. Farming Outlook 10(3):7–10

Shahid SA (2013a) Chapter 1: Developments in salinity assessment, modeling, mapping, and monitoring from regional to submicroscopic scales. In: Shahid SA, Abdelfattah MA, Taha FK (eds) Developments in soil salinity assessment and reclamation-innovative thinking and use of marginal soil and water resources in irrigated agriculture. Springer, pp 3–43

Shahid SA (2013b) Chapter 17: Food security constraints and role of biosaline agriculture in meeting the food demand in the Gulf States. In: Behnassi M, Olaf Pollman, Kissinger G (eds) Sustainable food security in the era of local and global environment change. Springer Science+Business Media, Dordrecht, pp 269–286

Shahid SA, Omar SAS (1999) Order 1 soil survey of the demonstration farm sites with proposed management. Kuwait Institute for Scientific Research, Kuwait, 144 pp. ISBN 0 957700369

Shahid SA, Rahman K (2011) Chapter 2: Soil salinity development, classification, assessment and management in irrigated agriculture. In: Pessarakli M (ed) Handbook of plant and crop stress CRC Press. Taylor and Francis Group Boca Raton London Newyork, pp 23–39

Shahid SA, Taha FK, Ismail S, Dakheel A, Abdelfattah MA (2011) Chapter 3: Turning adversity into advantage for food security through improving soil quality and providing production systems for saline lands: ICBA perspectives and approach. In: Behnassi M, Shahid SA, D'Silva J (eds) Sustainable agricultural development: recent approaches in resources management and environmentally-balanced production enhancement. Springer, Dordrecht/Heidelberg/New York/ London, pp 43–67

Shahid SA, Taha FK, Abdelfattah MA (eds) (2013a) Developments in soil classification, land use planning and policy implications-innovative thinking of soil inventory for land use planning and management of land resources. Springer, Dordrecht/Heidelberg/New York/London, 858 pp

Shahid SA, Abdelfattah MA, Othman Y, Kumar A, Taha FK, Kelley JA, Wilson MA (2013b) Chapter 1: Innovative thinking for sustainable use of terrestrial resources in Abu Dhabi Emirate through scientific soil inventory and policy development. In: Shahid SA, Taha FK, Abdelfattah MA (eds) Developments in soil classification, land use planning and policy implications-innovative thinking of soil inventory for land use planning and management of land resources. Springer, Dordrecht/Heidelberg/New York/London, pp 3–49

Shahid SA, Abdelfattah MA, Wilson MA, Kelley JA, Chiaretti JV (2014) United Arab Emirates keys to soil taxonomy. Springer, Dordrecht/Heidelberg/New York/London, pp xxxiv+108

Skwarnecki M, Fitzpatrick RW (2008) Geochemical dispersion in acid sulfate soils: implications for mineral exploration in the Mount Torrens-Strathalbyn area, South Australia. In: Fitzpatrick RW, Shand P (eds) Inland acid sulfate soil systems across Australia. CRC LEME Open file report no. 249 (Thematic volume). CRC LEME, Perth, pp 98–128

Soil Survey Staff (1999) Soil taxonomy: a basic system of soil classification for making and interpreting soil surveys, 2nd edn. USDA-Natural Resources Conservation Service, Agriculture handbook number 436. USDA-NRCS, US Government Printing Office Washington, DC 20402

Soil Survey Staff (2010) Keys to soil taxonomy. USDA-NRCS,11th edn

Tolba MK, Saab NW (2009) Arab environment climate change – impact on climate change on Arab countries. 2009 report of the Arab Forum for Environment and Development. Technical Publications and Environment & Development magazine, Beirut, 159 pp

von Braun J, Meinzen-Dick R (2009) Land grabbing by foreign investors in developing countries: risks and opportunities. IFPRI Policy Brief 13, Washington, DC, USA, April 2009, 9 pp

World Reference Base (2006) World reference base for soil resources: a framework for international classification, correlation and communication. IUSS Working Group WRB. World soil resources reports no. 103. ftp://ftp.fao.org/agl/agll/docs/wsrr103e.pdf

WWF-GFN (2012) Living planet report 2012. Published by WWF in collaboration with Global Footprint Network and Zoological Society of London, 164 pp. Avenue du Mont Blanc, Gland Switzerland

Chapter 2
Achieving Food Security in a Changing Climate: The Potential of Climate-Smart Agriculture

Mohamed Behnassi, Mohamed Boussaid, and R. Gopichandran

Abstract It's increasingly and widely recognized that the global warming rate is accelerating, exacerbated by humans' past and present unsustainable practices which result in human and environmental-induced effects and risks. Any move toward sustainable development models will involve significant paradigm shifts, particularly from the current economic models, in which it is presumed that a society can only develop by expanding its use of resources and increasing per capita consumption patterns, despite the related long term negative effects. In this context, and since climate change is generating risks and opportunities for agriculture, which is the main component of development strategies in many Southern countries, climate-smart agriculture (CSA) has recently emerged as a significant option in multilateral climate change debate. This option is believed to generate enormous benefits both in terms of adaptation, mitigation, and food security enhancement. This chapter aims at exploring how climate change is likely to impact agriculture, food production and security, and what actions can be taken to increase agriculture productivity, build resilience, and reduce GHG emissions through enhancing CSA, both in policies and practices. Given the inter-linkages between climate change, agriculture and food security policies, a governance approach has been recommended. Some of the

M. Behnassi (✉)
Global Environmental and Health Politics, Faculty of Law, Economics and Social Sciences, Ibn Zohr University of Agadir (Morocco), Agadir, Morocco

North-South Center for Social Sciences (NRCS), Agadir, Morocco
e-mail: behnassi@gmail.com

M. Boussaid
National Institute of Agronomic Research (INRA), Rabat, Morocco
e-mail: mohamed.boussaid@giz.de

R. Gopichandran
Environment and Climate Change Wing, Gujarat Energy Research and Management Institute, Gandhinagar, Gujarat, India
e-mail: r.gopichandran@germi.res.in

S.A. Shahid and M. Ahmed (eds.), *Environmental Cost and Face of Agriculture in the Gulf Cooperation Council Countries: Fostering Agriculture in the Context of Climate Change*, DOI 10.1007/978-3-319-05768-2_2, © Gulf Research Centre Cambridge 2014

most important guiding principles include equal emphasis on the management of natural resources, appropriate institutional and financial mechanisms, and improving preparedness of stakeholders to engage in well-informed actions.

Keywords Climate change • Food security • Climate-smart agriculture • Greenhouse gases • Governance

2.1 Introduction

This first decade of the third millennium has gained rapid international recognition, leading to global consensus that the global warming rate is accelerating, exacerbated by humans' past and present unsustainable practices. These practices, including those responsible for large volumes of greenhouse gas (GHG) emissions, deforestation, excess consumption of finite resources, reducing global biodiversity and contamination of water supplies, result in human-induced effects and risks that negatively impact on our quality of life. Politically, most countries agree that the debate about global warming is over, that climate change is a key symptom of how humans have impacted on planetary systems and that it is time for serious collaboration to help transform our institutional and individual practices, if next generations are to inherit a sustainable future. This concern and call for concerted action to tackle impacts of climate change, duly recognizing the environmental and related social problems, are reflected in the rising numbers of academic papers and popular literature articles since the 1960s. In general terms, transformation of global societies from mainly unsustainable practices to a more sustainable way of living will involve significant paradigm shifts, particularly from the current economic paradigm, in which it is presumed that a society can only develop by expanding its massive use of resources and increasing per capita consumption patterns, despite the long term negative effects of this behavior (Daly 1996).

In this context, many researchers, decision-makers, land use planners and civil society actors increasingly believe that the interaction between climate change and food security will be one of the biggest challenges for the coming decades. By the year 2025, 83 % of the expected global population of 8.5 billion will be living in developing countries, where most of the poor are living, and the resources are vulnerable to climate change. Yet the capacity of available resources and technologies to meet the demands of growing population remains uncertain. Presently, close to one billion people are already suffering from hunger worldwide and the future is daunting too: food needs are projected to increase by 70 % by 2050 when the global population reaches nine billion, while climate change is projected to reduce global average yields, among other severe consequences. Within this perspective, many believe that agriculture must become central to future climate-change and food security governance. This is on account of at least three important interrelated aspects:

- *Firstly*, agriculture is the sector most vulnerable to climate change and many threats, including the reduction of agricultural productivity, production stability and incomes in many areas of the world already characterized by high levels of food insecurity and limited means of coping with adverse climate impacts. Moreover, climate change will affect agriculture through higher temperatures, greater crop water demand, more variable rainfall and extreme climate events such as heat waves, floods and droughts. Many impact studies point to severe crop yield reductions in the next decades without strong adaptation measures, especially in areas where rural households are highly dependent on agriculture and farming systems are highly sensitive to inclement climate;

- *Secondly*, agriculture contributes a "significant" proportion of global carbon dioxide and nitrous oxide emissions (about 14 % of emissions according to current estimations) (Wright 2010); and
- *Thirdly*, agriculture can be a major part of the solution: helping people to feed themselves and adapt to changing conditions while mitigating impacts of climate change (carbon sequestration). This mitigation potential can be largely achieved in developing countries.

The need to tackle climate change while producing more food to feed the world's growing population means that "climate-smart agriculture" (CSA) is one of the advocated ways forward. This approach primarily defends an agriculture that sustainably increases productivity, resilience (adaptation), reduces/removes GHGs (mitigation). This will simultaneously help meet the goals of food security and overall development. This also envisions transformation of agriculture to feed a growing population in the face of a changing climate without corroding the natural resource base significantly and mitigate the negative effects of climate change. However, more productive and resilient agriculture will need better management of natural resources, such as land, water, soil and genetic resources through practices such as conservation agriculture, integrated pest management, agroforestry and sustainable diets.

This chapter aims at exploring as how climate change is likely to impact agriculture, food production and securities, and what actions can be taken to increase agriculture productivity, build resilience to tackle the negative impacts of climate change, and reduce GHG emissions through enhancing CSA – both in policies and practices. Given the inter-linkages between climate change, food security and agriculture policies, a governance approach has been recommended in this chapter. Some of the most important guiding principles include equal emphasis on the management of natural resources, appropriate institutional and financial mechanisms and improving preparedness of stakeholders to engage in well-informed actions.

2.2 Agriculture at the Intersection of Climate Change, Food Security, and Poverty Alleviation

As mentioned above, climate change is one of the main challenges facing our globalized world today since the science clearly indicates that a global temperature rise of 2 °C above pre-industrial levels may change the face of the world irreversibly. Furthermore, this challenge is increasingly considered as a 'threat multiplier' since it increases a range of livelihood threats and vulnerabilities, rather than being an isolated specific risk (IFAD 2010).

The poor population in developing countries will be particularly impacted by this delicate environmental problem, of which developed – and currently emergent – countries are the major drivers. In addition, food security, poverty and climate change are closely linked challenges and should not be considered separately. In countries where the economic and human development strategies are heavily based on agriculture, the development of agricultural sector, with a clear redistribution potential, remains an efficient poverty reduction policy. Yet agricultural expansion for food production and economic development which comes at the expense of soil, water, biodiversity or forests, environment, conflicts with other global and national goals, often compromises production and sound development in the longer term.

It is true that over the past six decades world agriculture has become considerably more efficient, especially in the 1960s through green revolution. Improvements in production systems and crop and livestock breeding programs have resulted in a doubling of food production while increasing the amount of agricultural land by 10 %. However, projections based on population growth and food consumption indicate that agricultural production will need to increase substantially to meet future demands. Most estimates also indicate that climate change is likely to reduce agricultural productivity, production stability and incomes in some areas already suffering from food insecurity, high rates of poverty, and feeble adaptive capacities to cope with adverse climate impacts. Preliminary estimates for the period up to 2080 suggest a decline of some 15–30 % of agricultural productivity in the most climate change-exposed developing country regions – Africa and South Asia (UNCTAD 2011). Hence, climate change is expected to exacerbate and multiply the existing challenges faced by agriculture and human security.

It is also true that human societies, over the centuries, have developed the capacity to adapt farming practices to environmental change and climate variability. These adaptations include, among others, practicing shifting cultivation, adopting high yielding, and new crop varieties tolerant to salts and drought and modifying grazing patterns. But today the speed and intensity of climate change are outpacing autonomous actions and threaten the ability of poor smallholders and rural societies to cope. For most of the one billion extremely poor and hungry people who live in the rural areas of major developing countries, agriculture remains the principal income source. These people are already vulnerable, and climate change will in most cases deepen their vulnerability. More specifically, and in countries most reliant on rainfed agriculture and natural resources, poor rural women, who are often the primary food producers, have fewer assets and less decision-making power, are even more exposed than men (IFAD 2010).

Therefore, ensuring food security under a changing climate should be considered as one of the major challenges of our era, especially that many countries' agriculture is highly vulnerable to negative impacts of climate change. Even using optimistic lower-end projections of temperature rise, climate change may reduce crop yields by 10–20 % by the 2050s, with more severe losses in some regions (Jones and Thornton 2009). World food prices for some of the main grain crops are likely to rise sharply in the first half of the twenty-first century, unlike the price declines witnessed in the twentieth century. Projections of price rises range from about 30 % for rice to over 100 % for maize, with about half or more than half of this rise due to climate change. Under a pessimistic high-end projection of temperature rise, the impacts on productivity and prices are even greater. Moreover, increasing frequencies of heat stress, drought and flooding events, not factored into the projections mentioned above, will result in further deleterious effects on productivity. It is likely that price and yield volatility will continue to rise as extreme weather continues. Climate change will also impact agriculture through effects on pests and disease. These interactions are complex and the full implications in terms of productivity are still uncertain (Gornall et al. 2010).

While agriculture is the most vulnerable sector, it is also a major cause of climate change, directly accounting for about 14 % of GHG emissions, and indirectly much more as agriculture is an important driver of deforestation and land-use change responsible for another 18 % of global emissions (IPCC 2007). Even if emissions in all other sectors were eliminated by 2050, growth in agricultural emissions in a business-as-usual (BAU) scenario, world with a near doubling in food production would perpetuate climate change. Therefore, while trying to cope with the effects of a changing climate, agriculture is simultaneously facing two other challenges: increasing food production in developing countries to meet population increases and dietary changes whilst remaining central to mitigation efforts (IFAD 2010).

2.3 Climate-Smart Agriculture: A 'Triple Win' Approach

A range of mitigation solutions is needed to tackle impacts and reduce the buildup of GHGs with implications on the 2 °C limit. The need for "no regret measures", and more precisely a truly sustainable and climate-friendly agricultural development, is currently less controversial than before. A glance at global mitigation potential shows that changes in agriculture and land use, including deforestation in tropical areas, presently account for one-third of global GHG emissions. Increasingly, therefore, agriculture is being recognized as part of the problem in global climate governance. While developed countries' emissions result mostly from industry, energy consumption and transport, the Food and Agricultural Organization figures reveal that 74 % of all agricultural emissions originate in developing countries, and 70 % of the agricultural mitigation potential can be realized in these same countries (Gattinger et al. 2011). For example, about half of the 47 African countries that have recently submitted Nationally Appropriate Mitigation Action (NAMAs) have included agriculture-related actions (Streck and Burns 2011). In general terms, agriculture has much to contribute to a low emissions development strategy. It can mostly contribute to mitigation (Smith et al. 2008) in three ways: avoiding further deforestation and conversion of

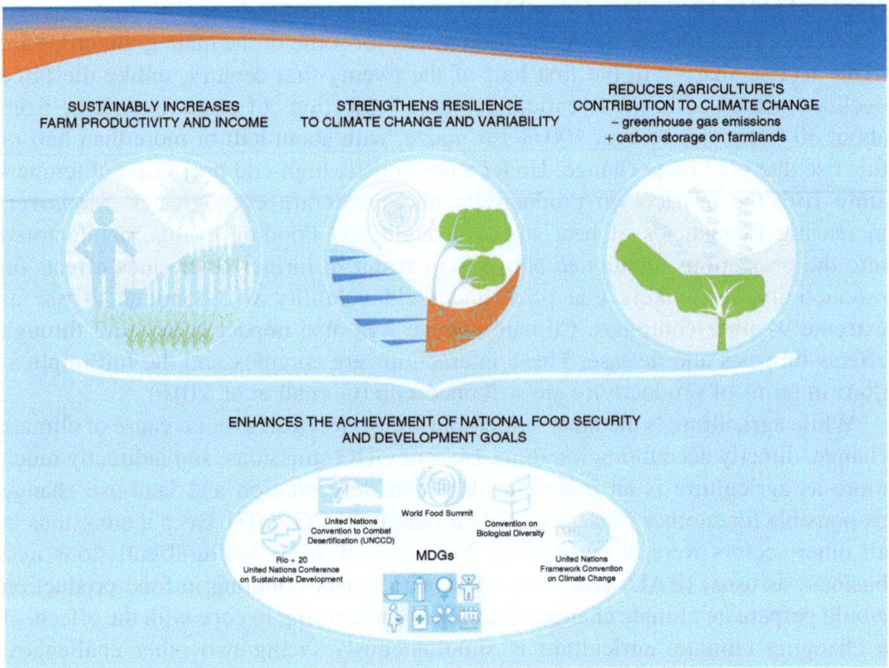

Fig. 2.1 Potential of climate-smart agriculture (CSA) (Source: FAO 2010)

grasslands and wetlands; increasing the carbon sequestration by vegetation and soil; and reducing current, and avoiding future, increases in emissions from nitrous oxide (from fertilizer use and soil organic matter breakdown) and from methane (from livestock production and rice cultivation) through appropriate cross-cutting and mutually enforcing policies, plans, programs and local initiatives.

Could agriculture therefore be part of the solution, particularly in developing countries? Globally, three-quarters of all malnourished people depend on agriculture and would be directly affected by international mitigation agreements aimed at agriculture. Various "climate- friendly" agricultural solutions have already been proposed. They include CSA, which has been advocated during the last climate negotiations in Durban (2011) as instrumental in achieving many aims.

CSA can be defined as an approach which seeks to increase productivity in an environmentally and socially sustainable way, strengthen farmers' resilience to climate change, and reduce agriculture's contribution to climate change by reducing GHG emissions and increasing carbon storage on farmland (Fig. 2.1). Climate-smart agriculture includes proven practical techniques – such as mulching, intercropping, conservation agriculture, crop rotation, integrated crop-livestock management, agroforestry, improved grazing, and improved water management – but also innovative practices such as better weather forecasting, early warning systems and risk insurance. It is about getting existing technologies off the shelf and into the hands of farmers and developing new technologies such as drought or flood tolerant crops to meet the demands of the changing climate. It is also about creating and enabling policy environment for adaptation (World Bank 2011).

The CSA approach neatly combines the twin goals of today's climate negotiators, helping to prevent climate change while at the same time adapting farms to inevitable change. It incorporates practices that increase productivity, efficiency, resilience, adaptive capacity, and mitigation potential of production systems (i.e. carbon sequestration). However, CSA requires more careful adjustment of agricultural practices to natural conditions, a knowledge-intensive approach, huge financial investment, and policy and institutional innovation, etc.

2.3.1 Climate-Smart Agricultural Production Systems: Relevant Practices

The production, processing and marketing of agricultural goods are central to food security and economic growth. Products derived from plants and animals include foods, fibers, fuels and raw materials and inputs for other industries. Production has been achieved through a number of production systems which range from smallholders mixed cropping and livestock systems to intensive farming practices such as large monocultures and intensive livestock rearing. The overall efficiency, resilience, adaptive capacity and mitigation potential of the production systems can be enhanced through improving its various components, some of the key ones are highlighted here to illustrate the feasibility and constraints of developing CSA. Two important manifestations of such interventions could be reduced deforestation and lesser encroachment of land systems for agriculture purposes. Other key issues, such as access to markets, inputs, knowledge, finances and issues related to land tenure are also fundamental for ensuring food security. These issues are also highlighted here:

2.3.1.1 Soil and Nutrient Management

The availability of nitrogen and other nutrients is essential to increase yields. This can be done through composting manure and crop residues, more precise matching of nutrients with plant needs, controlled release and deep placement technologies or using legumes for natural nitrogen fixation. Using methods and practices that increase organic nutrients inputs, retention and use are therefore fundamental and reduce the need for synthetic fertilizers which, due to cost and access, are often unavailable to smallholders and, through their production and transport, contribute to GHG emissions.

2.3.1.2 Water Harvesting and Use

Improved water harvesting and retention and water-use efficiency are fundamental for increasing production and addressing increasing irregularity of rainfall patterns. Today, irrigation is practiced on 20 % of the agricultural land in developing countries but can generate 130 % more yields than rain-fed systems. The expansion of efficient management technologies and methods, especially those relevant to smallholders, is fundamental.

2.3.1.3 Pest and Disease Control

There is evidence that climate change is altering the distribution, incidence and intensity of animal and plant pests and diseases as well as invasive alien species. The recent emergence in several regions of multi-virulent, aggressive strains of wheat yellow rust adapted to high temperatures is a good indication of the risks associated with pathogen adaptation to climate change. These new aggressive strains have spread at unprecedented speed in five continents resulting in epidemics in new cropping areas, previously not favorable for yellow rust and where well-adapted, resistant varieties are not available.

2.3.1.4 Resilient Ecosystems

Improving ecosystem management and biodiversity can provide a number of ecosystem services, which can lead to more resilient, productive and sustainable systems that may also contribute to reducing or removing GHG. Services include, control of pests and disease, regulation of microclimate, decomposition of wastes, regulating nutrients cycles and crop pollination. Enabling and enhancing the provision of such services can be achieved through the adoption of different natural resource management and production practices.

2.3.1.5 Genetic Resources

Genetic make-up determines a plants and animals tolerance to shocks such as temperature extremes, salts, drought, flooding and pests and diseases. It also regulates the length of growing season/production cycle and the response to inputs such as fertilizer, water and feed. The preservation of genetic resources (establishing gene banks, genetic engineering) of crops and breeds and their wild relatives is therefore fundamental in developing resilience to shocks, improving the efficient use of resources, shortening production cycles and generating higher yields per area of land. Generating varieties and breeds, which are tailored to ecosystems and the needs of farmers, is crucial.

2.3.1.6 Harvesting, Processing and Supply Chains

Efficient harvesting and early transformation of agricultural produce can reduce post-harvest losses and preserve food quantity, quality and nutritional value of the product. It also ensures better use of co-products and by-products, either as feed for livestock, to produce renewable energy in integrated systems or to improve soil fertility. As supply chains become longer and more complex, it becomes ever more important to increase the operational efficiency of processing, packaging, storage, transport, etc. to ensure increased shelf life, retain quality and reduce carbon footprints. Food processing allows surplus to be stored for low production years and allows a staggered sale. This ensures availability of food and income throughout the season and in years and low production. Food processing creates job and income opportunities, especially for women.

2.4 Transition Toward CSA: A Governance Approach

It is increasingly believed that the mitigation of climate change and the enhancement of food security will not be fully achieved without the transformation of agricultural systems toward more productivity, less variability in crop yields, and minimum carbon footprint. Such transformations can be supported by the adoption of climate-friendly practices as mentioned above, many of which not only improve food security but also deliver both reduced emissions and adaptation benefits – the "triple win". Careful selection of production systems, adoption of appropriate methods and practices and use of suitable varieties and breeds, soil management can allow considerable improvements to be made (Table 2.1). Currently, there are several resources, guidelines, tools (i.e. the Carbon Balance Tool developed by FAO to provide ex-ante estimations of the impact of agriculture and forestry development projects on GHG emissions and carbon sequestration, indicating its effects on the carbon balance) and other applications to assist policy makers, extension workers and farmers in selecting the most appropriate production systems, undertaking land use and resource assessments, evaluating vulnerability, and undertaking impact assessments (Bockel et al. 2012).

However, there are numerous and substantial challenges in transitioning to high production, intensified, resilient, sustainable and low-emission agriculture: i.e. there are still considerable gaps relating to the suitability and use of these production systems and practices across a wide variety of agro-ecological and socio-economic contexts and scales. There is even less knowledge on the suitability of different systems under varying future climate change scenarios and other biotic and abiotic stresses. In many cases, even existing knowledge, technologies and inputs have not reached all farmers (weak research-extension-farmer link), especially in developing countries. For this to be achieved there is a need for a governance system with the potential to provide inclusive policies, infrastructures

Table 2.1 Selected climate smart agriculture (CSA) best practices

Crops	Livestock	Agroforestry	Water management	Soil management	Fisheries and aquaculture
No tillage, direct seeding	Increased feeding efficiency, reducing methane emission through improving digestibility	Multipurpose trees on farms	Water harvesting, storage-e.g. water pans	Conservation agriculture	Saline resistant species
Rotations with legumes	Improved rangeland management	Nitrogen-fixing trees, bushes, fodder	Alternate wetting and drying (rice) rather ponding	Stone bunds	Increased feeding efficiency
Intercropping with legumes	Efficient treatment of manure	Improved fallows	Dams, pits, retaining ridges	Planting pits (zai)	Integration of aquaculture in farms
New varieties: shorter cycle, drought and salt tolerant etc.	Improved livestock health	Hedges, windbreaks, shelterbelts, live fences	Improved irrigation practices	Mulching, nutrient management	Low energy fuel efficient fishing
Improved storage and processing technologies	Animal husbandry improvements	Fruit orchards	Increased water productivity (crop per drop)	Reclaimed salt-affected lands	–

and considerable investments to build the financial and technical capacity of farmers (especially smallholders) while enabling them to adopt climate-smart practices which could generate sustainable rural livelihood, economic growth and ensure sustainable food security.

2.4.1 Institutional and Policy Options

Ensuring food security and development under climate change will involve increasing yields, income and production, which can generally be expected to lead to increased aggregate emissions. While agricultural production systems will be expected first and foremost to increase productivity and resilience to support food security, there is also the potential for enhancing low emission development trajectories without compromising development and food security.

To meet these multiple challenges, it has been suggested that a major transformation of the agriculture sector will be necessary and this will require institutional and policy support. Better aligned policy approaches across agricultural, environmental and financial boundaries and innovative institutional arrangements to promote their implementation will be needed. This section covers summarily the required critical adjustments to support the shift toward CSA:

2.4.1.1 Enabling an Integrated Policy Environment

Key requirements for an enabling policy environment to promote climate-friendly agricultural transformations are greater coherence, coordination and integration between climate change, agricultural development and food security processes. Inter-sectoral approaches and consistent policies across these areas are necessary at all levels. Such policies have both impacts on smallholder production systems and on GHG emissions. Lack of coherence can prevent synergy capture and render the pursuit of the stated policy objectives ineffective and costly.

- *At the national level*, climate change policies are generally expressed through the National Action Plan for Adaptation (NAPAs) and the Nationally Appropriate Mitigation Actions (NAMAs) as well as through national and regional strategies. Agriculture and food security plans are generally expressed in national development and poverty reduction strategies. Better alignment of the technology approaches envisioned in these different policy frameworks, and in particular better integration of sustainable land and water management factors into mainstream agricultural development planning will facilitate a more holistic approach to considering agricultural development, adaptation and mitigation. In addition, better integration of food security nets and adaptation policies offers the potential to reap significant benefits. Better use of climate science information in assessing risks and vulnerability and then developing the safety nets and insurance products as an effective response is already being piloted in some areas with fairly positive results. Policies related to price stability are also key to both adaptation and food security.

- *At the international level*, better integration of food security, agricultural development and climate change policies and financing is also needed. Two parallel global dialogues on reducing food insecurity and responding to climate change have until now had little substantive integration of issues under consideration. Likewise, the agriculture community has only recently become active in the discussions and negotiations of international climate change policies that could have profound impacts on the sector. The creation of mechanisms that allow dialogues between food security, agricultural development and climate change policy-makers seems fundamental and imposing.

2.4.1.2 Reducing Information Gap by Boosting Its Production and Dissemination

One key role of institutions is the production and dissemination of information, ranging from production and marketing conditions to the development of regulations and standards. The scientific and policy uncertainty pertaining to the management of climate change impacts as well as the costs of inaction, increase the value of information and the importance of institutions that generate and disseminate it. Thus, it will be critical that national and international agricultural research programs focus increasingly on developing countries to incorporate climate change into their policy-making processes.

In addition, improving the use of climate science data for agricultural planning can reduce the uncertainties generated by climate change, improve early warning systems for drought, flood, and pest and disease incidence and thus increase the capacity of farmers and agricultural planners to allocate resources effectively and reduce risks. Enhancing communication between producers and users of climate science is also clearly a requirement. Institutions to facilitate this exchange can be existing communications and information networks, including extension. Providing translation of climate information to planners and communities can also bridge the divide between science and field application. Capacity building of policy makers as well as technical staff is another avenue. Finally, platforms for collaborative action and information sharing which unites modelers, practitioners and donors, can enhance the development and use of climate science information for agricultural decision-making (FAO 2010).

The imperative of climate change also requires increased capacity of farmers to make both short and long term planning decisions and technology choices. Agricultural extension systems are the main conduit for disseminating the information required to make such changes. Yet in many developing countries, these systems have long been in decline due to weak linkage between research-extension and farmers. Resources have been severely curtailed and services increasingly outsourced to the private sector or dropped. Problems with delivering at a relevant spatial and time scale, difficulty in communicating the information and lack of user participation in development of information systems are all problems that have been encountered. It is equally important to consolidate information and insights from traditional knowledge of communities. These can be useful entry points for collective action strategies and enable a bottom–up approach.

2.4.1.3 Improving Access, Coordination and Collective Action

Input supply (i.e. access to fertilizers and seeds) is an activity that requires coordination beyond the farm. Given the market failures that lead to socially suboptimal use of seed and fertilizer, governments frequently step in to distribute them directly. Government-led distribution programs have often increased input use, but the fiscal and administrative costs are usually high and the performance erratic. Yet, cutbacks have often simply resulted in leaving smallholders without reliable access to seed and fertilizer. Producer organizations may offer a promising avenue to improving input supplies to smallholders. It is however important to ensure institutional arrangements of monitoring and verification of the quality of seeds and easy access to adapted varieties to guarantee environmentally sound management of productivity.

Many of the biophysical improvements to increase resilience in smallholder agricultural production systems identified above require action and coordination amongst many stakeholders in the rural landscape. Restoration of degraded areas to improve soil quality, better management of community water resources, and informal seed systems to facilitate the exchange of plant genetic resources are all examples of collective resource management activities that are likely to become more important in the climate change context. In many cases, local institutions exist to govern collective pressure due to population growth, conflicts, changes in market patterns and state intervention (FAO 2010). It will be useful to establish synergies amongst them for collective action.

2.5 CSA: Financing Mechanisms

A good starting point will be to mainstream agriculture in climate change policy making at different levels. This trend will enable agriculture to benefit from mobilized financial resources dedicated to support mitigation and adaptation actions. In addition, investments in CSA must link finance opportunities from public and private sectors and integrate climate finance into sustainable development agendas (FAO 2012). In general terms, significant finance, both public and private, is crucially needed if CSA is to be scaled effectively. Climate finance presents an opportunity with regard to this perspective, through carbon markets, performance-based donor finance for mitigation and adaptation, and private sector finance for agricultural production.

Already some retailers and manufacturers are supporting CSA by purchasing agricultural produce approved to standards for mitigation and adaptation. But further progress depends on recognition of the vital role smallholders can play in greening supply chains and enhancing resilience to climate change.

Future climate negotiations present an opportunity for governments to commit funds for CSA and to signal support for developing countries to set out NAMAs for agriculture. Carbon markets are another potential source of finance but reforms are needed if this option is going to work for agriculture. A Recent research (PWC 2012) found that if carbon markets are to fulfill their potential for supporting the scaling up of CSA activities, three key changes are needed:

- A wider range of CSA activities needs to become eligible in both compliance and voluntary carbon markets;
- More methodologies are needed that support 'triple-win' CSA practices; and
- The technical burden of CSA carbon project development needs to be reduced.

It's noteworthy that according to the World Bank (2011) new funds have been developed to increase food security, to respond to the food price crisis, to promote climate-resilient development, to reduce deforestation and forest degradation, or to support climate adaptation and mitigation more generally. In addition, the volume of finance associated with carbon markets is expanding rapidly. While a number of existing financing mechanisms have been instrumental in supporting climate change mitigation and adaptation, the FAO (2009) has indicated that the main mechanisms have generally not enabled agriculture to contribute fully to adaptation and mitigation efforts. The challenge for countries is to bring different funding mechanisms together so as to invest at the scale needed to achieve the goals of CSA. Practices that are profitable and self-sustaining in the long-run may need upfront finance to get off the ground. Capacity needs to be strengthened to enable concerned developing countries to access these existing and emerging climate finance mechanisms. There is also scope for the redirection of agricultural finance in developed and developing countries as well as development finance.

Patterns of public support which focus on research, investments in soil and water conservation, social protection and safety nets to enhance human capital and technology and value chain development are more effective, benefit more farmers and are more sustainable in the long run than price support (World Bank 2010). In China for example, investments in watershed management through public work programs based on food assistance have enabled impressive productivity increases. In Burkina Faso, investments in soil and water management from diverse stakeholders have powered what has been termed a "farming miracle" (FAO 2009). Participatory approaches directly involving farmers in decision-making generally work best. A key lesson is that the quality of public expenditure is as important as its quantity in facilitating private farmer investment in climate-smart agriculture.

2.6 Conclusions

The CSA offers some unique opportunities to tackle food security, adaptation and mitigation objectives. Many developing countries including Gulf States will specifically benefit from this option given the central role of agriculture in their economic and social development and the major negative impacts that climate change may cause them. Early action in CSA, while the global action continue with regard to mitigation, adaptation and food insecurity alleviation, is essential to build capacity, experience and guide future choices. The international community needs to demonstrate commitment to the multiple agendas of food security, adaptation and mitigation by stepping up investment support to CSA, in particular the scaling up of best practices and technologies as part of early actions. The chapter highlights some major aspects that have to be addressed in order to optimize the transition toward CSA while ensuring highly relevant outputs and outcomes. These have to be considered on a location and system specific basis and cannot be broad brushed across locations. Most importantly there is significant scope to enrich linkages across sectors including management of land, water and bio-resources and these have to be addressed on a priority basis and within a governance perspective.

References

Bockel L, Touchemoulin O, Jönsson M, Gorin P (2012) Is FAO's ex-ante carbon balance tool (EX-ACT) beneficial to the private sector? EASYPol module 117, FAO. Available at. http://www.fao.org/docs/up/easypol/907/EX-ACT_private-sector-117EN.pdf

Daly H (1996) Beyond growth. Beacon, Boston

Food and Agriculture Organization (2009) The investment imperative, paper from the FAO high level conference on world food security: the challenges of climate change and bioenergy. FAO, Rome

Food and Agriculture Organization (2010) Climate-smart agriculture: policies, practices and financing for food security, adaptation and mitigation. FAO, Rome

Food and Agriculture Organization (2012) Climate-smart agriculture: managing ecosystems for sustainable livelihoods002E FAO. Available at: http://www.fao.org/docrep/015/an177e/an177e00.pdf

Gattinger A, Jawtusch J, Muller A, Mäder P (2011) No-till agriculture – a climate smart solution. http://m.misereor.de/fileadmin/redaktion/MISEREOR_no%20till.pdf

Gornall J, Betts R, Burke E, Clark R, Camp J, Willett K, Wiltshire A (2010) Implications of climate change for agricultural productivity in the early twenty-first century. Phil Trans R Soc B 365:2973–2989. doi:10.1098/rstb.2010.0158

IFAD (2010) Climate change strategy: enabling poor rural people to overcome poverty. http://www.ifad.org/climate/strategy/e.pdf

Intergovernmental Panel on Climate Change (IPCC) (2007) Climate change 2007: synthesis report. Contribution of working groups I, II and III to the fourth assessment report of the intergovernmental panel on climate change. IPCC, Geneva

Jones P, Thornton P (2009) Croppers to livestock keepers: livelihood transitions to 2050 in Africa due to climate change. Environ Sci Policy 12:427–443

Price Waterhouse Coopers (PWC) (2012) Making climate finance work for smallholder famers in sub Saharan Africa/agricultural carbon markets: opportunities and challenges for sub Saharan Africa, PWC. Available at: http://www.pwc.co.uk/assets/pdf/making-climate-finance-work-for-small holder-farmers-in-sub-saharan-africapdf.pdf

Smith P, Martino D, Cai Z, Gwary D, Janzen H, Kumar P, McCarl B, Ogle S, O'Mara F, Rice C, Scholes B, Sirotenko O, Howden M, McAllister T, Pan G, Romanenkov V, Schneider U, Towprayoon S, Wattenbach M, Smith J (2008) Greenhouse gas mitigation in agriculture. Philos Trans R Soc B 363:789–813

Streck C, Burns D (2011) Towards policies for climate change mitigation: incentives and benefits for smallholder farmers. CGIAR Research Program on Climate Change, Agriculture and Food Security (CCAFS), Frederiksberg

UNCTAD (2011) Assuring food security in developing countries under the challenges of climate change: key trade and development issues of a fundamental transformation of agriculture. http://archive.unctad.org/templates/page.asp?intItemID=2101&lang=1

World Bank (2010) The Hague conference on agriculture, food security and climate change: opportunities and challenges for a converging agenda: country examples. World Bank, Washington, DC

World Bank (2011) Opportunities and challenges for climate-smart agriculture in Africa, World Bank. http://climatechange.worldbank.org/sites/default/files/documents/CSA_Policy_Brief_web.pdf

Wright Julia (2010) Feeding nine billion in a low emissions economy: challenging, but possible – a review of the literature for the overseas development institute. Oxfam/ODI

Chapter 3
Prospects of Agriculture in the State of Kuwait-Constraints and Opportunities

Habibah S. Al-Menaie

Abstract Kuwait is one of the Gulf Cooperation Council (GCC) member countries. Kuwait is water scarce country. The agriculture sector contributes only 0.4 % of the country's Gross Domestic Product (GDP). The engagement of local people to agriculture is less than 1.1 % and it is projected that climate change will greatly impact agricultural sector of Kuwait. Kuwait is one of the least agricultural countries in the world and the arable land amounts to less than 9 % of total acreage. Ninety one percent of national water requirements are reliant on costly desalinated water and 54 % of the total water is used for productive agriculture. Hence, there is an urgent need for adapting sustainable and economical crop production system to enhance production efficiency, productivity and quality. To overcome this situation, the Kuwait Institute for Scientific Research (KISR) aims to incorporate applied research into integrated farming systems, sustainable crop production and animal production technologies. This chapter focuses on constraints and possibilities of agriculture in Kuwait and on the major agricultural production research works conducted at Aridland Agricultural Production (AAPP), and Biodiversity for Terrestrial Ecosystems (BTEP) Programs of KISR.

Keywords Biocapacity • Desert • Ecological footprint • Harsh climatic conditions • Integrated farming systems

3.1 Introduction

Kuwait is one of the Gulf Cooperation Council (GCC) member countries and is located in the north–eastern part of the Arabian Peninsula. It occupies an area of about 17,818 km^2 in which almost 90 % is a desert. With the continued rapid

H.S. Al-Menaie (✉)
Environment and Life Science Center, Kuwait Institute for Scientific Research,
PO Box 24885, Safat 13109 Kuwait
e-mail: hmanaie@kisr.edu.kw

S.A. Shahid and M. Ahmed (eds.), *Environmental Cost and Face of Agriculture in the Gulf Cooperation Council Countries: Fostering Agriculture in the Context of Climate Change*,
DOI 10.1007/978-3-319-05768-2_3,

expansion of oil sector, diversification of economy and constraints to local agriculture production, the agriculture sector contributes only 0.4 % of the country's Gross Domestic Product (GDP). Being poor and scarce in terrestrial resources, hyper-arid (precipitation/potential evapotranspiration <0.05) and under critical water stress ($>10{,}000$ persons per Million Cubic Meter of water) conditions, Kuwait's Ecological Footprint of Consumption is 9.72 global hectare per person (gha) compared to biocapacity of 0.43 gha per person, with a nominal deficit of 9.29 gha per person (WWF-GFN 2012) (Kuwait uses more resources than locally generated), the deficit is compensated through extensive food import. Kuwait being capital rich nation due to rich oil resources, it has no foreign exchange limitation for food import. The engagement of local peoples to agriculture is less than 1.1 % and it is projected that climate change will greatly impact agricultural sector of Kuwait. Kuwait is one of the least agricultural countries in the world and the arable land amounts to less than 9 % of total acreage. The United Nations (UN) classifies Kuwait as being exposed to extreme water scarcity. Ninety one per cent of national water requirements is reliant on costly desalinated water and 54 % of the total water is used for productive agriculture. Hence, there is an urgent need for adapting sustainable and economical crop production system to enhance production efficiency, productivity and quality. To overcome this situation, the Kuwait Institute for Scientific Research (KISR) aims to incorporate applied research into integrated farming systems, sustainable crop production and animal production technologies. The emphasis is given on efficient use of available land, water, energy and plants in agriculture. In the new strategic plan, KISR also aims to develop modern agricultural technology to improve production performance of promising crops such as forage barley and wheat for achieving food security. The crop water use efficiency can be maximized through the promotion of modern irrigation systems and the introduction of suitable crop varieties. This chapter focuses on constraints and possibilities of agriculture in Kuwait and on the major agricultural production research works conducted at Environment and Life Science Center of KISR. The continuation of these researches is vital in improving the agricultural prospects of Kuwait.

3.2 Climatic Conditions

The climate of Kuwait can be classified as hyper-arid to arid as per the Atlas of Desertification (Middleton and Thomas 1997). According to Le Houerou (1992), Kuwait and northern part of Saudi Arabia cover the arid areas of the GCC. Climate of Kuwait is characterized by extremely hot dry summers with long intense sunshine and moderately cool short winters with occasional rain. Summer is long (May–September) and hot with a mean temperature of 40 °C and a maximum temperature of about 50 °C. Winter is short with long annual average rainfall of about 120 mm. The rainfall, which occurs between mid October and late April, is minimal with an average of 115 mm year^{-1}. The rainfall pattern of Kuwait follows that of arid climates, where it is both sparse and unpredictable to create a reliable source of irrigation water. Evaporation is very high, ranging from 3 to 16 mm day^{-1}. Low relative humidity and strong dry and hot northwesterly winds prevail during summer, particularly in June and July. Sand storms are of common occurrence during spring, summer and fall. According to Brown and Al-Mazrooei (2003), dust storm affects the life of the inhabitants within the country, and mostly due to overgrazing and human interruption of the natural environment.

3.3 Soil Types

Soils of Kuwait are sandy and poorly developed, low in organic matter and water retention capacity, high level of salts and poor soil forming reaction. Aridsols (70.8 %) and Entisols (29.2 %) are the main soil orders of Kuwait (Roy and Omar 2005; Omar and Shahid 2013). The top horizon of the soil has sandy or sandy loam texture since these soils are developed mostly from whitish calcareous sandstones or gravelly sand that is cemented with lime. A hard pan layer locally known as "*gatch*" is found at various depths, which restricts root penetration and deep-water percolation leading to the formation of local water tables. The major primary minerals are sodium feldspar, plagioclase and orthoclase. Clay fraction contains montmorillonite and vermiculite type clay minerals. Infiltration rate of the desert soil is normally very high ranging from 15 to 100 cm h^{-1} (Abdal and Suleiman 2007). According to Misak et al. (2002), land degradation processes prevail in about 70 % of the terrestrial environment in Kuwait.

Soil information of a particular area is essential for the scientists as well as the farming community for sustainable farming. Soil survey revealed that there are 785,000 ha of land with soils potentially suitable for irrigation development in Kuwait, out of which 23,000 ha are currently being used for agriculture. A further 177,000 ha are alienated from agricultural development because of restrictive land uses and/or higher value land uses. Hence the challenge for scientists, planners and decision makers is now to delineate areas with suitable soils for irrigation development. It is also important to develop strategies and policies for irrigation in accordance with the targets set in the Agricultural Master Plan.

Field survey and laboratory analysis confirmed that the soil types at Ash-Shiquaya and Al-Abdally farms were of the Aridisols soil order while at Al -Wafra they were Entisols. Soils at Al-Abdally generally contain high proportions of $CaCO_3$ and equivalents and gypsum ($CaSO_4 \cdot 2H_2O$), and surface crusting is a problem in hindering the rate of seedling emergence. Application of organic mulches and polymers to these soils is suggested to decrease crust formation. High $CaCO_3$ limits the availability of micronutrients, but management practices which reduce the pH and make it more acid may reverse the trend. Those sites with very deep hard pan may be used to grow shallow rooting plants using increased fertilizer rates. At Al-Wafra the soils are sandy and low in fine particles and organic matter. Water loss is high due to rapid infiltration through the soil and perched water tables are possible with irrigation where hard pan layers are present. Wind erosion is likely because of the loose and sandy nature of the soils, so shelter breaks around the farm are suggested (KISR 1999a, b).

In Al-Abdali, an area of 50 ha was designated to develop a demonstration farm. The first order survey of this area revealed different types of soils and hence the area can be used for irrigated agriculture as well as for animal husbandry which will lead to long-term sustainable development (Shahid and Omar 1999; Shahid et al. 2004). Since the soil is poor in organic matter, the soil can be made fertile through the incorporation of organic materials and green manure (Shahid et al. 2004). In addition, the soil survey conducted at Al-Wafra area showed that though the soils are non-gypsiferous, calcareous, non-saline and low in organic matter, these can be used for cultivation through proper management (Al-Menaie and Al-Shatti 2006).

3.4 Irrigation Water Resources

Kuwait is a desert country with no surface waters. The only natural water resource in Kuwait is ground water and UN classifies Kuwait as being exposed to extreme water scarcity. Since there is nearly no natural source of fresh water, 91 % national requirements is reliant on costly desalinated water and 54 % of the total water is used for productive agriculture and landscaping projects. In Kuwait, irrigation is essential for any agricultural activity since it is part of the arid and semi-arid region. Even though 200,000 ha of land has the potential for irrigated agriculture, less than 3 % is currently under cultivation due to the shortage of irrigation water and extent of soil salinity (KISR 1999a, b). Irrigated land, although small in area, accounts for a large proportion of crop production. Although irrigation leads to phenomenal enhancement of land productivity, mismanagement of this resource will cause soil degradation and increased salinity. Fresh water is provided by desalination of sea water, which is then blended with 10 % aquifer brackish water. Another source of water available for agricultural use is treated municipal wastewater (1,300–2,450 mg L^{-1} Total Dissolved Solids (TDS)). Ground water is the only natural water resource in Kuwait that can be used directly for agriculture without pretreatment. However, fresh water (600–1,000 mg L^{-1} TDS) supplies are small and represent only 0.4 % of ground water reserves, the bulk of Kuwait's ground water is brackish. A total of about 100 million cubic meters (mM^3) of brackish water is produced annually from various fields throughout the country. The groundwater is used for landscaping, agriculture and in the oil industry. It has been estimated that utilization of groundwater resources at current rates will result in the severe depletion of the resources in 50 years.

To meet the increasing demand for irrigation water in Kuwait, the Sulaibiya Wastewater Treatment and Reclamation Plant was initiated in March 2005, which treats around 60 % of Kuwait's total domestic wastewater using the reverse osmosis (RO) technique in domestic wastewater reclamation. The Plant initially treated up to 375,000 cubic meter of raw domestic wastewater per day and would eventually expand its capacity up to 600,000 cubic meters per day during the 30-year concession period. When fully operational, the Plant is believed to contribute 26 % of Kuwait's overall water demand, reducing the annual demand from non-potable sources from 142 to 26 mM^3. Even though the specifications of the Reclaimed Water (RW) produced from the Plant exceeds the World Health Organization (WHO) standards for potable water, its output is used only for irrigation. The main objective of this Plant is to recharge the reclaimed water into the underground aquifer so as to serve as a strategic water reserve for the State of Kuwait. In addition to this, other benefits include prevention of marine environmental pollution since only partially treated wastewater would be disposed off in the Gulf, and transform sludge into natural fertilizer suitable for agricultural purposes, and thereby save the Government from the financial burden of huge investments. The government of Kuwait heavily subsidizes water production. The production cost of 1,000 imperial gallons (IG) (4,545 m^3) of water for the Ministry of Electricity and Water (MEW) is KD 3.855 (US$14.26), while the customer is charged only KD 0.8 (US$2.96) for it. Treated wastewater produced from Sulaibya plant is supplied to farmers @ KD 0.200 (US$0.74) for 1000 IG. Sulaibya plant treats wastewater with the additional processes of ultrafiltration (UF) and Reverse Osmosis (RO), and hence, it costs an additional amount of KD 0.500 (US$1.85).

3.5 Constraints in Agriculture Sector

Agriculture areas in Kuwait include Wafra, Abdali, Sulaibya, City and Greater Greenbelt area (Fig. 3.1). At present agriculture in Kuwait is constrained by many factors including physical, climatological technological, manpower and economical factors. Physical constraints comprise mainly the water and soil effects. The Kuwait soil is sandy in texture having low water retention capacity, low cation-exchange-capacity (CEC), low organic matter, low water holding capacity, poor soil forming reaction and low in available phosphorus. Also the sandy soil is having a very high infiltration rate of 15–100 cm h^{-1} and gatch layer was found at various depths.

Water Resources in Kuwait are very scarce since there are no surface waters which are considered a major constraint in agriculture production. Kuwait is hyper-arid (precipitation/potential evapotranspiration <0.05) and is under critical water stress ($>$10,000 persons per mM^3 of water). Ground water is the only natural water resource which is used directly for agriculture without pretreatment, and fresh water (600–1,000 mg L^{-1} TDS) represents only 0.4 % of ground water reserves. Brackish water and treated waste water are the other two sources for irrigation.

It is also estimated that climate change will greatly impact agriculture sector of Kuwait. The harsh climatic conditions entail extended period of high summer temperatures, high winds and dust storms, little or no rainfall, short growing season,

Fig. 3.1 Map showing locations of agriculture areas in Kuwait

limited natural water resources, high soil and irrigation water salinity, greater frequency of drought, increased insect and pest problems, fragile ecosystem, high evaporation rates, sand and dust storms and lack of sufficient natural greenery which make Kuwait one of the least agricultural countries in the world. Arable land amounts to less than 9 % of total acreage. Degradation of land is another factor affecting sustainable development due to overgrazing of rangelands by livestock, military activities, camping and recreation activities, these constraints become obvious during periods of extended drought. Kuwait has suffered severe land degradation in recent decades (Khalaf 1989; Omar 1990; Brown 2003). According to Zaman (1997), grazed areas contained 53 % less plant cover and 3.3 times higher bare grounds than protected areas.

Technological constraints include lack of improved varieties tolerant to drought and salinity. Availability of improved crop production technology and its adoption remains limited in Kuwait. Conventional plant production operations have caused economic and ecological problems associated with a short growing season, increased costs of energy based inputs, declining farm incomes, soil and water pollution and soil erosion. Lack of market intelligence and stiff competition from imported produce is another limitation. Agriculture scenario in Kuwait is also constrained by manpower factors such as lack of technological expertise and lack of skilled laborers.

With the continued rapid expansion of oil sector, diversification of economy and the constraints to local agriculture production, the agriculture sector contributes only 0.4 % of the country's Gross Domestic Product (GDP). To overcome these problems, a variety of measures need to be undertaken so as to utilize the desert landscape in a sustainable manner.

According to FAO (2003), food security exists when all people, at all times, have physical, social, and economic access to sufficient, safe, and nutritious food that meets their dietary needs and food preferences for an active and healthy life. In view of the fact that Kuwait is one of the most arid regions in the world; it faces enormous food security and food sovereignty challenges and is dependent on cereal and non-cereal food imports. Kuwait being capital rich nation due to rich oil resources, it has no foreign exchange limitation for food import or in foreign investment in agriculture sector to improve its food security. The expenses incurred for local food production is more expensive compared to importing. Moreover, desertification being a global threat to food security, combined efforts of multidisciplinary teams of researchers in national, regional and global organizations, in association with communities and policy makers is essential to provide solutions to combat desertification. Furthermore, due to many constraints to local agricultural production, the engagement of local people in agriculture is less than 1.1 %.

Even though Kuwait is limited by these factors, an important aspiration of the country like other countries in the Arabian Peninsula is to achieve at least a modest level of self-sufficiency in food production. The government has experimented in conducting research for high-tech agriculture which has been achieved by using advanced greenhouses and control room facilities, advanced biotechnological options, growing food through hydroponics and carefully managed farms. Since farmers always tend to reduce the cost of cultivation, they will refuse to adopt costly new technologies. At this juncture, it is the duty of the research institutes to adopt sophisticated innovative agricultural technologies to augment crop production. Since the water sources are insufficient, research on the use of drought tolerant lines is a priority area of study.

Despite the several challenges facing the agriculture sector in Kuwait, high priority is given to the development, intensification and modernization of this sector (food, livestock, greenery and fisheries). According to the data on ecological footprint (Global Footprint Network 2010), the world-average ecological footprint in 2007 was 2.7 global hectares per person with a world-average biocapacity of 1.8 global hectares per person. Kuwait's Ecological Footprint of Consumption is 9.72 gha per person compared to biocapacity of 0.43 gha per person. Since, Kuwait does not have enough ecological resources within its own territory, there is a local ecological deficit and hence it is an ecological debtor country with an ecological remainder value of −9.29 and the deficit is compensated through extensive food import.

3.6 Opportunities and Future Perspectives for Agriculture

Farming is being practiced in an area of 46,965 ha (2.71 %) in Kuwait. The future expansion of agriculture sector in Kuwait is guided by the Agricultural Master Plan (1995–2015), with a major emphasis on sustainable utilization of available land and water resources in agriculture (Roy and Grealish 2004). Future perspectives include use of improved irrigation practices, use of drought tolerant crops, water harvesting techniques, protected agriculture, integrated pest management conservation and use of native genetic resources and, application of biotechnology. The limitation of crop production in Kuwait due to high summer temperatures and water scarcity can be overcome by the use of cooled greenhouses and winter season crops. Crop yield varies considerably due to many factors but production per unit of water is much greater from protected greenhouse and plastic tunnel crops than those grown in the field.

The development-plan policies of agriculture sector of Kuwait for the period 2010–2014 seeks to cause tangible changes in the growth rates of gross agricultural product and accordingly, increasing the percentage of its contribution in the GDP.

The policies include:

- Targeting the growth of the agricultural product with a rate of 20.1 % annually through encouraging the private sector to invest in this sector and thereby increasing its percentage of contribution in the GDP.
- Encouraging the investment in the different agricultural fields (plant – animal – fish – beautification) and giving higher role for the private sector to establish the projects of agricultural production and marketing.
- Creating national cadres specialized in all agricultural fields, and making the local courses for developing and polishing the technical skills of the individuals working in the sectors of the Public Authority for Agriculture and Fisheries in cooperation with the Arabian, regional and international organizations.
- Establishing a geographical information system (GIS) for the agricultural plot, and connecting it with the systems of support available for the national product (plant – animal – fish), and working on developing the methods of distribution and censorship.
- Developing and updating the systems and mechanisms of agricultural marketing in cooperation with the private sector, and providing an integrated automatic system for the agricultural information and an electronic database, and working on entering into bilateral agreements with the countries advanced in the agricultural field and effectuating the agreements signed by the friendly and brother countries.

It is a fact that there is not a common platform for sharing the knowledge gained from the scientific research among the six member states of the Gulf Cooperation Council. Whenever such opportunities arise, Kuwait is attempting to share the acquired knowledge and technological innovations. If there is an established network between the Arab countries, it will be easier to collaborate with regional experts for the dissemination of knowledge. Kuwait is always willing to share its scientific knowledge free of charge.

The major challenge of the agricultural sector in Kuwait is to maximize land and water productivity without degrading the environment and natural resource base. To overcome this obstacle KISR, being the pioneer institute in agricultural research, aims to develop efficient farming systems, evolving crop and animal production technologies to conserve and sustainably utilize natural resources and also to enhance greenery under Kuwait's harsh climatic conditions. For this, the emphasis is given on sustainable and efficient use of available land, water and energy. KISR also aims to select and adapt crop plants that can tolerate increasing salinity levels in soil and tolerate continuous irrigation with poor quality water. Hence, in the new strategic plan, KISR is also planning to develop modern agricultural technology to improve production performance of promising crops such as forage barley and wheat for achieving food security. The crop water use efficiency can be maximized through the promotion of modern irrigation systems such as sprinklers, center pivot, drip, bubbler system etc., replacing crops with high water requirements with drought tolerant crops and using treated sewage effluent irrigation. The efforts in the area also include applied research into integrated farming systems, sustainable crop production and animal production technologies. These modern technologies will serve the needs of the local agricultural sector in increasing economical and efficient agricultural production.

The poultry industry in Kuwait which forms part of the agriculture sector is one of the leading food industries in Kuwait and is moving towards production of specialty food such as eggs enriched with omega-3-fatty acids which will take part of the market share in the future. Since, local producers supplied only 47 % of the poultry meat and 55 % of table eggs consumed and the remainder of the local poultry consumption imported from other countries, more development and improvement in production is needed in this industry (Al-Nasser 2006).

KISR has accepted the fact that for agriculture to expand in Kuwait's environment, major efforts must be made to select and adapt plants that can tolerate heat and salinity on the one hand, and to exploit the potentials of those plants with inherent tolerance on the other hand. To achieve this end, KISR had prepared a comprehensive "Conceptual and Strategic Framework for Biosaline and Biothermal Program Development in Kuwait" which highlighted the work that was done over the long-term and an Operational Plan to help to progress the nation agriculture forward.

The major agricultural research works done at KISR in the Environment and Life Science Center include:

- Introduction of flowering trees of the genus Cassia for the enhancement of greenery in Kuwait, Phase I: Introduction and evaluation (Al-Menaie et al. 2009)
- Selection of crops for salt tolerance (Abdal et al. 2002)
- Evaluation and improvement in *Zizyphus* for landscape beautification and fruit production in Kuwait (Bhat et al. 2004).
- Selection of salt-tolerant ornamental plants (Suleiman et al. 2005)
- Use of hydrophylic polymers and mulches for water conservation in greenery projects in Kuwait (Bhat et al. 2006)

- Standardized methods for the production of high quality gardenia (*Gardenia jasminoides*) in Kuwait (Al-Menaie and Al-Shatti 2007)
- Selection of olive varieties for greenery and fruit production in Kuwait: Phase I (Bhat et al. 2008)
- Selection of drought-tolerant ornamental plants

3.6.1 Introduction of Flowering Trees of the Genus Cassia for the Enhancement of Greenery in Kuwait, Phase I: Introduction and Evaluation (Al-Menaie et al. 2009)

The KISR has achieved remarkable success in introducing new flowering ornamental trees such as *Cassia fistula* and *Cassia nodosa* for aesthetic beautification in Kuwait. Various seed germination experiments were conducted to find out an efficient protocol on seed germination and the results revealed that mechanical scarification using sandpaper yielded 100 % germination and proved that manual scarification was the most effective way of breaking physical dormancy of *C. nodosa* and *C. fistula* seeds. Plant growth assessment was also conducted by studying the effect of soil mixture, fertilizer concentration and the interaction of soil mixtures with fertilizer concentrations. To study the effect of soil mixture, three soil mixtures at 1:1:1, 2:1:1 and 3:1:1 ratios of sand: peat-moss: humus was tested. The study revealed that the plants grown in 1:1:1 ratio recorded a higher plant height and higher number of leaves compared to 2:1:1 and 3:1:1 for both *C. nodosa* and *C. fistula*. The study on the effect of fertilizer concentrations using three fertilization regimes (N:P:K @ 1, 2 and 3 g L^{-1}) showed that seedling growth response was more pronounced in N: P: K @ 1 g L^{-1} in both the species. Results of the interaction effect of sand: peat-moss: potting soil at ratios of 1:1:1, 2:1:1 and 3:1:1 with three fertilizer regimes (N: P: K @ 1, 2 and 3 g L^{-1}) revealed that soil mixture in 1:1:1 ratio with 1 g L^{-1} of N: P: K was the best for *C. nodosa* and *C. fistula*.

3.6.2 Selection of Crops for Salt Tolerance (Abdal et al. 2002)

To enhance Kuwait's crop production and to increase crop productivity KISR initiated a project to screen salt-tolerant crops. To achieve these objectives, a screening study was conducted using potential salt tolerant genotypes collected from various international research organizations. The study was conducted under laboratory conditions using a nutrient solution recommended by the US Salinity Laboratory. The seedlings thus obtained from the germination studies were grown in plastic containers filled with perlite under greenhouse conditions. The seedlings were irrigated using salinized nutrient solution of various concentrations (1.6, 9.6, and 18.6 dS m^{-1}).

The whole study was based on a systematic reduction in the investigated genotype at various controlled stages, namely, germination, seedling establishment and growth to adult size. At each stage, those varieties that exhibited maximum ability to withstand increasing salinity treatments were passed on to the later stages of experimentation for further screening. The crops that were found to be the most tolerant to salinity were grown under field conditions in Kuwait using brackish water, and freshwater was used as a control. At the end of the study, five crops were selected as being the most tolerant to saline water in Kuwait which includes Tashkentskaya 10 for cabbage, Motano for cauliflower, Detroit Dark Red 1 for beetroot, IG33000 for barley, Bl Bhsto De Chinchon for garlic, and IP 3757 for pearl millet.

3.6.3 Evaluation and Improvement in Zizyphus *for Landscape Beautification and Fruit Production in Kuwait (Bhat et al. 2004)*

To evaluate the growth performance of improved cultivars of *Ziziyphus* and to investigate opportunities for their large scale use in greenery development and fruit production under Kuwait's environmental conditions, a 3 year project was conducted at KISR. During the first year, local *Zizyphus* plantations were surveyed and 35 promising lines were selected and propagated. A field study was conducted using 24 improved varieties and 35 selected local lines during the second and third years of the project. All varieties established well under coastal conditions, and after 24 months of planting produced an average height ranging between 312.8 cm (Thailand Selection) and 97 cm (Pakistan Selection 3) and the average increase in stem girth was between 2.2 (Pakistan Selection 3) and 4.6 mm (Pakistan Seedless). The study showed that the varieties Pakistan Seedless, Gola, Khalsi, Kuwait selection 1 and UAE selection can perform well under Kuwait's environmental conditions.

Another study was conducted to evaluate the response of ten improved varieties to irrigation with varying levels of salinity. The grafts were planted in 15-gal plastic containers filled with agricultural soil and irrigated with freshwater until they were established. Salinity stress was administered by irrigating the plants with salinized nutrient solution containing different amounts of total dissolved solids concentrations (5, 10 or 20 dS m^{-1}). The study revealed that the varieties differed in their response to salinity with some withstanding total dissolved salts (TDS) of 12,800 mg L^{-1} in the irrigation water without any significant reduction in plant height, while others were affected by TDS above 6,400 mg L^{-1}. On the contrary, greenery impact of these varieties were adversely affected at the highest salinity level (12,800 mg L^{-1} TDS).

3.6.4 Selection of Salt-Tolerant Ornamental Plants (Suleiman et al. 2005)

To determine the response of ornamental plants to saline irrigation water, experimental plants were planted in 15-gallon plastic containers filled with agricultural sand and irrigated with salinized nutrient solution containing different amounts of total dissolved salt concentrations (1,000, 1,600 or 3,200 ppm TDS) was used. A complete randomized block design with three replications was used. The plant height of five salt tolerant plant species under different salinity treatments was recorded. Based on preliminary data, *Allamanda cathartica*, *Duranta goldiana*, *Peltophorum ferruginum*, *Ficus pumila* and *Thespisia populnea* appeared to be highly tolerant to salinity. Growth in other species was only mildly affected by highest salinity (5.0 dS m^{-1}) treatment (Suleiman et al. 2005).

3.6.5 Use of Hydrophylic Polymers and Mulches for Water Conservation in Greenery Projects in Kuwait (Bhat et al. 2006)

To improve the water-use-efficiency in greenery projects and to conserve available water resources, a project was initiated to assess the potential for application of hydrophilic polymers and mulches for water conservation in greenery plantations under harsh, arid climatic conditions of Kuwait. In order to achieve these objectives five tasks were undertaken, namely, selection of polymer to improve water holding capacity of sandy soil, effects of temperature regimes on the effectiveness of polymers, effects of polymer-mulch combinations on plant growth, greenery impact and water conservation in selected ornamental plants, effect of polymer on growth and water-use in lawn grasses and demonstration of the most suitable polymer and mulch combination in an established landscape garden.

Of the five polyacrylamide polymers tested under greenhouse conditions, Agrihope @ 0.4 % was found to be the best in conserving water without any diverse effects on plant growth and soil properties. These polymers were found to be effective when fresh water was used for irrigation under all three temperature regimes (variable ambient, greenhouse and uniform laboratory). In general, polymers were far less effective under the variable ambient conditions, whereas the plants grew slowly and consumed least amount of water under the uniform laboratory temperature. Since Agrihope @ 0.4 % reduced plant water requirement to the maximum extent (33 %), it was selected to use in field trials. In the first field trial, Agrihope was mixed with agricultural sand @ 0.4 % for backfilling the pits after planting. Two each of the trees (*Conocarpus lancifolius*, *Zizyphus spinachristi*), shrubs (*Bougainvillea glabra, Hibiscus rosa-chinensis*) and ground covers (*Alternenthra versicolor, Gazania uniflora*) were used. After planting, one of the two mulches (horticulture grade bark or gravel) was spread on the soil surface. Results obtained during the first 12 months after planting suggested that polymer and organic (bark) mulch combination effectively reduced the daily water requirements of these plants between 19 and 50 %. Polymer effects were significantly reduced by time. The polymer and mulch did not have any impact either on greenery qualities of these plants or on physical and chemical soil parameters.

In another field trial, two lawn grasses (*Paspalum vaginatum and Zoysa japonica*) were established in the soil that was amended with 0.4 % Agrihope. Application of polymer reduced the water requirements by 25–29 % in early and by 7 and 15.9 % in later stages of plant development in *Paspalum* and *Zoysa* respectively. Thus, the studies conducted established the benefits of incorporating Agrihope @ 0.4 % into the soil before planting and then applying 5 cm thick layer of bark mulch to the soil surface after planting.

3.6.6 *Standardized Methods for the Production of High Quality Gardenia (***Gardenia jasminoides***) in Kuwait (Al-Menaie and Al-Shatti 2007)*

This study dealt with gardenia (*Gardenia jasminoides*), a shrubby evergreen plant with fragrant flowers and handsome foliage, and the work focused on testing standardized horticultural and management practices which leads to growing gardenia successfully in Kuwait. Ten replicates of indoor variety and five replicates of outdoor variety were used in two separate experiments. They were planted in five gallon plastic containers filled with sand alone and soil alone as two different treatments and another six treatments were prepared by using a mixture of potting soil, agricultural sand, perlite and peat moss in the ratio 1:1. Data on number of branches, leaf area index, chlorophyll content, canopy and number of flowers were recorded to evaluate the performance of the plants and found that canopy; chlorophyll and leaf area index were significantly influenced by these treatments, but flowering was least affected of the various treatments used, soil: peat moss in the ratio 1:1 was found to be the best followed by soil: perlite in both the varieties.

3.6.7 *Selection of Olive Varieties for Greenery and Fruit Production in Kuwait (Bhat et al. 2008)*

Containerized plants of ten olive cultivars (*Istanbuli, Frantoio, Koroneiki, Black Italian, Barnea, UC13A6, Leccino, Picual, Corotina* and *Arbequina*) were used to ascertain their response to irrigation water salinity. For this, response of ten cultivars to irrigation water salinity (1.6, 5, 10 or 20 dS m^{-1}) was assessed in 15-gal containers under the shade house environment.

The results indicated cultivar Barnea to be highly tolerant salinity levels up to 10 dS m^{-1}, whereas cultivars, Arbequina, Corotina, Istanbuli, Koroneiki and Picual were found to be moderately tolerant and Black Italian, Frantoio, Leccino and UC13A6 to be susceptible to irrigation water salinity.

An irrigation task was conducted to determine the response of selected olive cultivars to induced water stress at various stages of development with a view to develop an efficient irrigation schedule based on actual requirement at different stages of tree development. Five cultivars namely Barnea, Coratina, Arbequina, Koroneiki and UC13A6 were tested at three levels of irrigation (50, 75, 100 % ETo). Data on vegetative parameters such as plant height, stem diameter, number of branches, time of resprouting, rate of growth, number of nodes and inter nodal length showed that none of the varieties was adversely affected by even the highest water stress level (50 % of ETo) indicating that these varieties are drought-tolerant under Kuwait's environmental conditions.

3.6.8 Selection of Drought-Tolerant Ornamental Plants

Salinity and drought are now the major determinants of global plant production as millions of acres of arable land are lost from production each year due to these causes. While salinization is the occurrence of excessive soluble salts in soils, drought is the absence of rainfall or irrigation for a period of time sufficient to deplete soil moisture and injure plants. Based on their response to water stress, plants are classified as drought sensitive or tolerant plants. Drought symptoms resemble salt stress because high concentrations of salts in the root zone also cause water loss from roots. Salt- and drought plants maintain their turgor at low water potential by increasing the number of solute molecules in the cell.

According to Omar and Bhat (2008), Kuwait's native vegetation provides a valuable indicator of human perturbation, besides offering valuable gene pool and plant material for drought and salt-tolerance research. To enhance the sustainability of farming operations and to conserve local biodiversity, KISR had initiated a research program on organic agriculture in an effort to make the agricultural sector more eco-friendly and sustainable.

3.7 Conclusions

The agricultural sector of Kuwait is facing many challenges including the inadequacy of water resources necessary for production process and the dependability of the cultivated space. Even though the economy of Kuwait is developing in a fast pace, with the high income from oil, increase in population leads to the limitation of agriculture sector to fulfill the food requirements of the nation. Also the prevailing climate of Kuwait makes agricultural production relatively more expensive and difficult. The extreme climatic and soil conditions coupled with low rainfall adversely affect crop productivity. Further, the effect of climate change worsens the challenges of dry areas which are characterized by acute water scarcity and land degradation. Since, climatic challenges impose constraints on sustainable agricultural development, greater emphasis is needed to safeguard natural resources and agro-ecological practices.

In spite of these challenges, advances in science and technology, and closer cooperation and partnerships between various organizations provide numerous opportunities. In addition, practices involving integrated water and land management, sustainable crop production systems; crop improvement including improved irrigation practices, water harvesting techniques, introduction of drought tolerant crops, conservation and use of indigenous genetic resources, and application of biotechnology helps in sustainable agricultural development.

Since, the state relies extensively on imported food from other countries, leading to food insecurity, there is an urgent need for adapting sustainable and economically viable crop production system to enhance production efficiency, productivity and quality. Hence, KISR aims to increase agricultural production in a sustainable manner through conservation agriculture technologies to help maintain the productivity of ecosystems. Developing new crop varieties adapted to climate change and tolerant to drought, heat and salinity as well as biotic stress is a key priority. Application of molecular biology and biotechnology offers great promise in developing such varieties. The projects conducted at KISR played a significant role in developing strategies for increasing the production and productivity of agricultural crops in Kuwait. Experience gained from this work and information obtained will be of great value in improving the status of agriculture in Kuwait. The results of the present studies and data may be useful for future studies. The continuation of these researches is crucial to improve environment and crop production with good quality and high yielding. In addition, development of linkages between KISR, Public Authority for Agriculture Affairs and Fish Resources (PAAFR), Farmers' Union and other organizations is pertinent these goals.

Acknowledgement The author thanks the management of the Kuwait Institute for Scientific Research and the Kuwait Foundation for the Advancement of Sciences for their continued interest, encouragement and financial support to this project.

References

Abdal MS, Suleiman MK (2007) Soil conservation through mulching. Desertification in arid lands: causes, consequences and mitigation. Paper presented at the international conference on desertification control in the arid region, 12–15 May 2007

Abdal MS, Bhat N, Albaho M, Al-Ghawas S, Ferin J, Ghloum D (2002) Selection of crops for salt tolerance. Final report, Kuwait Institute for Scientific Research, Kuwait

Al-Menaie H, Al-Shatti A (2006) Soil survey and site selection for crop cultivation in Kuwait. Cereal science and technology for feeding ten billion people: genomics era and beyond.

In: Proceedings of the meeting of the Eucarpia Cereal Section, Lleida, Spain, 13–17th Nov 2006, pp 463–465
Al-Menaie H, Al-Shatti A (2007) Pilot scale production of high quality gardenia (*Gardenia jasminoides*) in Kuwait. HS technical report. Kuwait Institute for Scientific Research, Kuwait
Al-Menaie H, Al-Ragam O, Al-Shatti A, Mathew M, Suresh N (2009) Introduction of flowering trees of the genus *Cassia* for the enhancement of greenery in Kuwait, Phase I: Introduction and evaluation. Final report. Kuwait Institute for Scientific Research, Kuwait
Al-Nasser A (2006) Poultry industry in Kuwait. World's Poult Sci J 62:702–708
Bhat NR, Suleiman MK, Al-Mulla L, Al-Menaie H, Bellen R, Sharma A, Al-Zalzaleh M (2004) Evaluation and improvement in *Zizyphus* for landscape beautification and fruit production in Kuwait. KISR 7284. Kuwait Institute for Scientific Research, Kuwait
Bhat N, Suleiman MK, Al-Menaie H, Al-Ali EH, Al-Mulla L, Al-Zalzaleh M, Bellen R (2006) Use of hydrophylic polymers and mulches for water conservation in greenery projects in Kuwait. Final report. Kuwait Institute for Scientific Research, Kuwait
Bhat NR, Suleiman MK, Al-Menaie H, Al-Mulla L, Al-Dosery S, D'Cruz G, Bellen R, Thomas B, Ali SI, George P, Al-Zalzaleh M (2008) Selection of olive varieties for greenery and fruit production in Kuwait: Phase I. Final report. Kuwait Institute for Scientific Research, Kuwait
Brown G (2003) Factors maintaining plant diversity in degraded areas of Northern Kuwait. J Arid Environ 54:183–194
Brown G, Al-Mazrooei S (2003) Vegetation regeneration in a seriously degraded *Rhanterium epapposum* community in Northern Kuwait after four years of protection. J Environ Manage 68:387–395
Food and Agricultural Organization (FAO) (2003) Rural women: key to food security. Gender and Population Division, Rome
Global Footprint Network (2010) Ecological Footprint Atlas 2010. Available at http://www.footprintnetwork.org/images/uploads/Ecological%20Footprint%20Atlas%202010.pdf. Retrieved 28 Oct 2010
Khalaf FI (1989) Desertification and aeolian processes in the Kuwait desert. J Arid Environ 16:125–145
KISR (1999a) Soil survey for the State of Kuwait, vol I. Executive summary. Kuwait Institute for Scientific Research and Public Authority for Agriculture and Fish Resources, AACM International Pty Ltd. Adelaide, 1:26–29
KISR (1999b) Soil survey for the State of Kuwait: reconnaissance survey, vol II. AACM International Pty Ltd., Adelaide
Le Houerou HN (1992) Man-made deserts: desertization processes and threats. Arid Land Res Manage 16:1–36
Middleton N, Thomas D (1997) World atlas of desertification. United Nations Environment Program/Arnold, 182 pp
Misak R, Al-Awadhi JM, Omar SAS, Shahid SA (2002) Soil degradation in Kabd area, Southwestern Kuwait city. Land Degrad Dev 13:403–415
Omar SAS (1990) Desertification in the Eastern Region of Arabian Peninsula. The case study of Kuwait. PhD dissertation, University Microfilms International, University of California, Berkeley, Ann Arbor, Michigan
Omar SAS, Bhat NR (2008) Alteration of the *Rhanterium epapposum* plant community in Kuwait and restoration measures. Int J Environ Stud 65:139–155
Omar SA, Shahid SA (2013) Chapter 3: Reconnaissance soil survey for the state of Kuwait. In: Shahid SA, Taha FK, Abdelfattah MA (eds) Developments in soil classification, land use planning and policy implications: innovative thinking of soil inventory for land use Planning and management of land resources. Springer, Dordrecht/Heidelberg/New York/London, pp 85–107
Roy W, Grealish G (2004) Mapping arable soils using GIS based soil information database in Kuwait. Manage Environ Qual Int J 15(3):229–237

Roy W, Omar SAS (2005) Application of GIS in soil classification and land use planning for the State of Kuwait. Available at www.kfupm.edu.sa. Accessed 14 Apr 2012

Shahid SA, Omar SAS (1999) Order 1 soil survey of demonstration farm sites with proposed management. Report no. KISR 5463. Kuwait Institute for Scientific Research, Kuwait

Shahid SA, Omar SAS, Jamal ME, Shihab A, Abo-Rizq H (2004) Soil survey for farm planning in Northern Kuwait. Kuwait J Sci Eng 31(1):43–57

Suleiman MK, Bhat NR, Al-Mulla L, Bellen R (2005) Evaluation of new ornamental plants for use in Kuwait's landscape demonstration garden establishment. Report no. KISR 6601. Kuwait Institute for Scientific Institute, Kuwait

Zaman S (1997) Effects of rainfall and grazing on vegetation yield and cover of two arid rangelands of Kuwait. Environ Conserv 24:344–350

Chapter 4
Improved Water Use Efficiency Is a Key Practice to Sustain Food Production in the Gulf Cooperation Council Countries

Asadullah Al Ajmi

Abstract Water scarcity in the Gulf Cooperation Council (GCC) region has already become a challenge to future agriculture development, especially open field. Water and arable lands are limited in the GCC countries. This is due to low rainfall and hot desert environment, resulting into 65–90 % consumption of water resources by agriculture sector. The rate of ground water abstraction is higher than the aquifer recharge leading to a decline of ground water level and increase in salinity through sea water intrusion. In addition agricultural sustainability is also hampered by harsh environmental conditions with evapotranspiration exceeding over ten folds than precipitation. Moreover, the region lacks surface water to compensate the imbalance of water deficit. This justifies that farming is only possible through ground water and treated waste water (TWW) irrigation using different irrigation systems. The latter offsets the water scarcity to some extent. The soils are generally sandy with high drainage capacity requiring improvement of water use efficiency through modern irrigation systems and modeling such as IMAGE – a physically-based non-intensive data acquisition model developed from data collected from small farms irrigated with saline water in Oman. Other option is to use a soilless hydroponic system that utilizes agro-technologies to improve water use efficiency. The Sultan Qaboos Center for Modern and Soilless Agriculture in the Arabian Gulf University Bahrain hosts strong data base helping the investors for science based project planning. The TWW is an assured resource and the only one that is guaranteed to increase in response to population growth.

Keywords GCC • WUE • Irrigation management • Modeling • Hydroponic • Soilless

---

A. Al Ajmi (✉)
Soils and Irrigation, Desert and Arid Zones Sciences Programme, College of Graduate Studies, Arabian Gulf University, Manama, PO Box 26671, Bahrain
e-mail: asadmohe@gmail.com

S.A. Shahid and M. Ahmed (eds.), *Environmental Cost and Face of Agriculture in the Gulf Cooperation Council Countries: Fostering Agriculture in the Context of Climate Change*, DOI 10.1007/978-3-319-05768-2_4, © Gulf Research Centre Cambridge 2014

4.1 Introduction

The Gulf Cooperation Council (GCC) countries are located in the desert environment, which is characterized by high temperatures, high evaporation rates and low and erratic rainfall. The territories of the member states of GCC countries (Bahrain, Kuwait, Oman, Qatar, Saudi Arabia, and the United Arab Emirates) occupy most of the Arabian Peninsula (Fig. 4.1); an area of huge reserves of crude oil and gas. Water scarcity in the GCC region has already become a challenge to future agriculture development, especially open field. Demand management, therefore, should be the first priority to effectively use available water resources in the agriculture sector. The average annual rainfall in the GCC countries varies between 70 and 140 mm (Al-Rashed and Sherif 2000). The total area of the GCC countries is 2,423,300 km^2, while the total population is estimated at about 42.1 million (GCC General Secretariat Official Website 2012). Perennial rivers and lakes do not exist in any of the GCC countries; therefore, surface water resources are scarce with the exception of the mountainous areas in southwestern part of Saudi Arabia, eastern part of United Arab Emirates and northern and southern parts of Oman. The groundwater resources can be classified into renewable resources, which are mostly encountered in small shallow alluvial aquifers, and non-renewable resources (or fossil water) which are encountered in the deep aquifers. Aquifer recharge

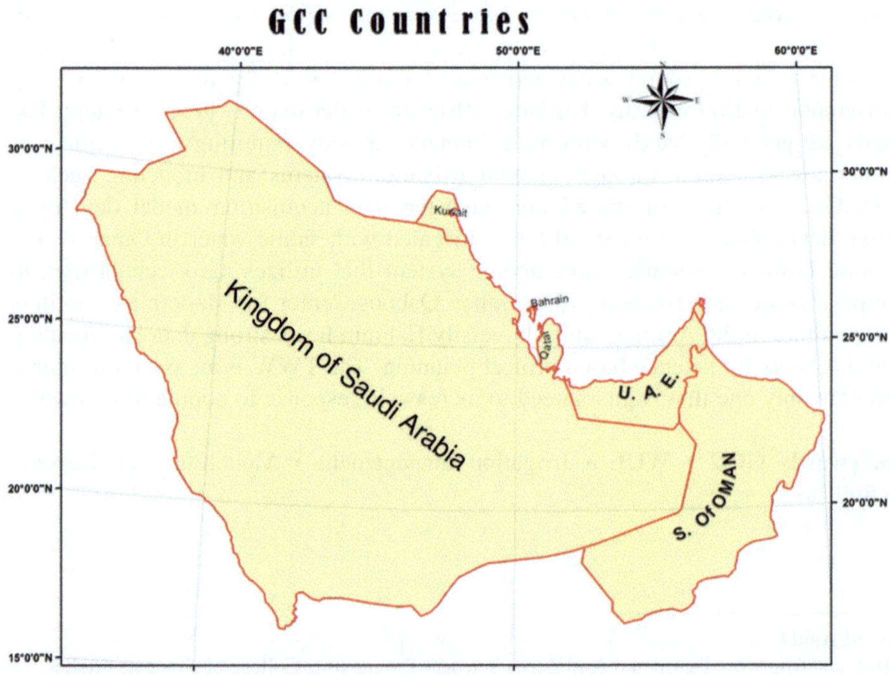

Fig. 4.1 The geographical location of the GCC countries (Source: AGU, GIS/RS DataBase)

depends on the rainfall events and surface runoff, and thus may vary considerably from 1 year to the other. The reliable groundwater reserves are encountered in the thick extensive sequences of the sedimentary formations of the Arabian Shelf underlying two-third of the Peninsula (Al-Rashed and Sherif 2000).

Oil and gas production is prime commerce of the region. The recent hike of international oil prices has raised the living standard and increased infrastructure development in the GCC countries.

The increase of local agriculture production and industrial activities has imposed additional stresses on the water demand. Urban landscapes, public parks and private gardens are taking good share of water sector. Over irrigation and leakage wastes of irrigation water thus needs careful management for better use of water resources. Emphasis has been given for water reuse, recycling and to recharge the depleting aquifers. Between 75 and 80 % of the water resources in the region are used by agriculture sector (Alsharhan et al. 2001), and World Bank (2005) reported such estimates as 78 %. Most of the agricultural water (85 %) is groundwater, which is largely non-renewable (Bazza 2005). The rate of ground water abstraction is higher than the recharge and thus aquifer levels are rapidly declining, and ground water salinity is increasing through sea water intrusion (World Bank 2005). This has led to a sharp decline in the availability of ground water for irrigation purpose.

Water scarcity in the GCC countries is seen as a big challenge limiting agricultural development (Zubari 1997). However, there are technological options to reduce water consumption and demand in an environmentally acceptable manner and without any appreciable impact on the lifestyles. These include minimizing unaccounted-for water, adopting a demand approach rather than supply approach, alternate water sources such as use of reclaimed or treated wastewater, and water conservation through localized irrigation systems, low-flow toilets, low-flow shower heads and faucet flow restrictors. These technologies have gained significant considerations owing to the risen cost of municipal water (Al Salem and Abouzaid 2006). In spite of the harsh environment conditions and severe shortage in water resources, water utilization in various sectors is far from being efficient and judicious, thus, future agriculture in the region is highly threatened and unsustainable. However, if opportunities are realized for reversing the situation before it is irreversible and strong policies of water management are established and backed up with effective legislations and procedures for implementation, the situation can be handled in a sustainable manner.

With the rapid increase (4.77 %) of population from 36.807×10^6 in 2006 to 38.564×10^6 in 2008, it is projected that the population of GCC countries will reach to 44×10^6 by 2020 (United Nations (UN) 2002). Water scarcity and high consumption rates in the GCC region has also accelerated the depletion of the poor groundwater resources. This can be justified, that water consumption in the United Arab Emirates is around 353 L per person a day compared to 420 L in the USA (United Nations (UN) 2002), which is more than three times the world average. This shows that the per capita renewable water resources in the GCC region are the lowest in the world. It is visualized that freshwater consumption in the GCC countries is likely to increase continuously as a result of high incomes, comfortable lifestyles, real-estate development, the availability of energy for desalination, and growth in the tourism industry. In contrast, regional agricultural water consumption is characterized by low productivity, and has been significantly affected by droughts in recent years (WWAP 2012). Therefore, there is an increase in demand coupled with the reliance on expensive means to provide water. In the last few years, oil revenues increased dramatically. By some estimate (McClatchy 2007), oil revenues tripled between 2002 and 2005, rising from 25 % of GDP to 38 %. With that increase in government revenue, the whole economic system received an impressive boost. Having limited renewable resources along with the increase in demand, most governments in the GCC region have devoted lots of financial resources to provide potable and clean water to its residents. The move towards utilizing the desalination option was intensified. It is known that these plants are expensive, energy intensive, impacting the environment (from green house gas (GHG)) emissions and concentrates from desalination process, in addition to their short economic life and expensive maintenance cost, nevertheless, this was agreed to be the best option for the region. Arabian Gulf countries rely heavily on desalination for freshwater resources. Saudi Arabia currently has the world's greatest desalination capacity, followed by the UAE as the second largest producer. Jointly, they produce more than 30 % of global freshwater production (ESCWA 2009). By some estimates, there is a need to invest US$100 billion over the next 10 years for new desalination plants in order to keep up with the 6 % increase in demand for the desalinated water (United Nations (UN) 2002). In the GCC region, desalination capacity must be doubled to over $5{,}000 \times 10^6$ gal a day to meet the projected demand for the year 2015. Based on 2007 unit costs, the GCC countries require about US$20 billion investment in desalination.

Considering all the constraints and challenges in water use and management in the GCC countries in mind, this chapter provides an overview of the current water use practices with special reference to agricultural sector and an insight to the potential areas for improvement.

4.2 Climate Change and Water Shortage in the MENA Region

Middle East and North Africa (MENA) region includes 21 countries including all GCC countries. World Bank (2012) recently reported that under average climate change scenario (CCS), MENA's will have water shortage five folds by the year 2050, from current (42 km^3) to 200 km^3 (2050). Demand gap is expected to vary from 85 km^3 under the wet CCS to approximately 283 km^3 under dry CCS. Moreover assuming the most likely average trend for CC, agricultural water demand will increase by approximate 25 % (ranging from a 15 % increase to a 33 % increase in irrigation water demand under the wetter and warmer climate trend, and the warmer and drier climate trend, respectively) (World Bank 2012). It is projected that the conditions in the GCC region will be even worse owing to unavailability of surface water and low rainfall. Currently the demand gap in GCC region is met through unsustainably mining fossil groundwater and expensive desalination water.

4.3 Improving Water Use Efficiency for Crop Production

Water use efficiency (WUE) refers to different processes and ratios in the literature. The term efficiency in general reflects the ratio of output to input. In water sector, this terminology is widely used in irrigation systems and engineering design and evaluation and management. Water conveyance efficiency, for instance, reflects losses of water from the distribution system. Water application efficiency refers to the losses of water below crop root zone. Water distribution efficiency is the uniformity of water across the irrigated field and water storage efficiency reflects the efficiency of water stored in the crop root zone (Hansen et al. 1980; Jensen 1981; Walker and Skogerboe 1987; James 1988; Keller and Bliesner 1990).

Since higher yield per unit of water used (crop per drop) is one of the most important challenges in dry land agriculture, Condon et al. (2002) used the term "intrinsic water-use efficiency" and W_T is referred to the ratio of the instantaneous rates of CO_2 assimilation and transpiration (T) at the stomata. They concluded that improving W_T may be one means of achieving this goal. Numerous studies have found highly variable relationship between crop yield and W_T. It should be remembered, that, the impact on crop yield of genotypic variation in W_T will depend on three factors: (i) the impact of variation in W_T on growth rate, (ii) the impact of variation in W_T on the rate of water use, and (iii) how growth and water use interact over the crop's duration to produce grain yield?. The relative importance of these three factors will differ depending on the crop species being grown and the nature of the cropping environment.

In this chapter WUE is referred to the ratio of yield to water used during crop growth. In the water-scarce regions such as the GCC region, water (not land) is the most limiting factor to crop production. Satisfying crop water requirements may maximize production from the land unit, but it does not necessarily maximize the return per unit volume of water. Therefore, improving water productivity can contribute to water savings, which can be used to irrigate additional lands with higher total production and/or improve the sustainability of the existing water resources. It is assumed that maximum WUE may be achieved at irrigation levels below those that satisfy full crop irrigation requirements.

4.4 Water Resources Management Approaches

All GCC countries have pursued supply-side approaches to address the increasing water demands. This has included dam building and reservoirs, water desalination reuse, and new technologies to improve the efficiency of conventional and non-conventional methods including water harvesting and efficient ways of water delivery. The large dams may have negative environmental and social impacts, but they help reduce the uncertainty and risk related to floods, droughts and climatic variabilities. For water resources to be sustainable in the GCC region, their management has to be shifted from a supply-oriented approach to a more demand-management approach. Large savings in water, energy and financial resources are expected as a result of increased efficiency and reduced consumption in the region. Continuing challenges facing the demand-management approach include evaluation of availability and demand in watersheds, possible reallocation or storage expansion in existing reservoirs, balancing equity and efficiency in water use, inadequate legislative and institutional frameworks, and the rising financial burden of aging water infrastructures. More efforts need to be placed towards reducing the water demand in various sectors in the region. There is a large potential for the region to improve WUE using success stories from some developing countries. As an instance, Singapore, reduced its urban domestic water demands from 176 L per capita per day in 1994, to 157 L per capita per day in 2007 (Kiang 2008). Leak detection programmes in Bangkok and Manila have lowered estimated unaccounted-for water, allowing new infrastructure development to be postponed (WWAP 2009).

In the GCC countries, the farming systems are dominated by small scale irrigated holdings, which are cultivated by conventional crops mainly date palms and fodders irrigated with ground water which is pumped into gravity-driven open earth channels leading to crops' irrigation basins.

Dominant soils of the region are sandy in nature (MAW 1985; MAF 1990; KISR 1999; Scheibert et al. 2005; EAD 2009a, b; Omar and Shahid 2013; Harahsheh et al. 2013; Shahid et al. 2013a, b), in mountainous areas these are gravelly, and hence present high percolation rates. Irrigation management is rarely based on the actual water demand by crops. In most cases, water losses through evaporation and deep percolation exceed by several folds the actual crop water needs for transpiration (Al-Ajmi et al. 2002). Several strategies have been suggested to improve the water productivity and management of water resources. The following sections address some practical oriented solutions to the water stress problem characterizing the region. The common feature among these solutions is to improve the efficiency by which water is applied to and used by crops. This chapter deals with agricultural water management aspects.

4.4.1 Transforming Current Irrigation Practices to Efficient Methods

The majority of the agricultural farms in the region use traditional water conveyance methods, application and management. Thus, the current efficiency of water application is very low due to unnecessary wastage. There is a large potential to improve the efficiency of currently used methods. Rehabilitation of traditional irrigation channels, conveying irrigation water through closed conduits, and the use of modern irrigation methods are just few examples to mention. Soils are generally sandy in texture with extensive rate of infiltration leading to appreciable amount of irrigation water to be lost through drainage and seepage through the conveyance channels. Lining irrigation channels with concrete can improve the conveyance efficiency to a large extent. Modern irrigation systems (drip, bubblers, sprinklers, subsurface) when used properly can save large amount of water. Modern irrigation techniques are essential for the GCC region to increase the WUE.

Drip irrigation (DI) allows uniform delivery of water and nutrients directly to the plant root zone, therefore can increase nutrient use efficiency compared with other irrigation methods (Sammis 1980; Miller et al. 1981). Surface and subsurface drip irrigation methods are the most effective ways to save water by efficiently using water and to increase the crop yield (Tiwari et al. 1998; Al-Omran et al. 2005). Many researchers (Hutmacher et al. 1996; Ayars et al. 1999) reported a significant increase in crop yield and WUE. The implementation of drip irrigation was identified as the best strategy in terms of water savings and application of water on farmlands followed by conveying irrigation water through closed conduits. Improved efficiencies through drip irrigation and improved water distribution systems will have demonstrable effects. Subsurface drip irrigation (SDI) is the latest advanced method of irrigation, which enables the application of a small amounts of water to the soil through the drippers buried below the soil surface with discharge rate generally in the same range of surface drip irrigation (ASAE Standards 1999). The SDI offers many advantages over DI (Camp 1998). Both surface and subsurface drip irrigation methods may be managed to decrease deep percolation and surface runoff of water (Phene et al. 1989; Lamm and Trooien 2003). Al-Omran et al. (2010) conducted a field experiment to investigate the effects of irrigation methods on the yield and water use of tomato in Saudi Arabia. Based on the results obtained, the study recommended SDI irrigation at 25 cm drip line depth with 80–100 % (ETc) irrigation level to obtain a good growth and yield of tomato under the same conditions. The DI has a high actual uniformity, but the soil water distribution in the soil region between the emitters is non-uniform in SDI method (Wallach 1990).

The effect of drip line depths on the growth and yield of tomato has been evaluated by various researchers. In loamy soils with germination assistance objectives, Charlesworth and Muirhead (2003) recommended a drip line depth of no greater than 20 cm. For row crops a drip line depth of 45 cm is recommended (Ayars et al. 1999) in Yolo clay loam soil, when tomato seeds were placed directly above the drip lines, better germination results were achieved with drip line depths of 15 cm or 23 cm (Schwankl et al. 1990). The rational irrigation can significantly increase yield (Gajre et al. 1997), excessive irrigation may lead to decrease crop WUE (Jin et al. 1999), while deficit irrigation may result in higher yield and definitely higher WUE (Demir et al. 2006). Deficit irrigation is one of maximizing

WUE for higher yields per unit of irrigation water applied: in the deficit irrigation system the crop is exposed to certain level of water deficit during a particular growth stage or throughout the whole growing season. Many studies have been conducted to determine water production function using evapotranspiration (ET) or water applied (W). The relationship between seasonal ET and yield (Y) showed a linear function (Musick et al. 1994), while the relationship between water applied and yield is curvilinear (Simsek et al. 2005; Sun et al. 2006). However, El-Shafei (1989) found a linear relationship between Y and W for tomato but up to the irrigation level that equals to 80 % of pan evaporation.

4.4.2 Use of Modeling Approach

Models can be utilized for many different aspects of water resources assessment and management, for planning and research. To meet the needs of these diverse applications, a range of models has been developed. These can range from simple models with few parameters to very complex models with many parameters. Selection of the appropriate model can have a great influence/impact on the modeling output and an inappropriate choice can generate misleading results with serious consequences for management decisions and resources planning (Al-Qurashi 2010).

4.4.2.1 Water Balance Models

Rainfall and evaporation are amongst the key variables necessary for modeling recent and current groundwater recharge in the Arabian Peninsula (Scanlon et al. 2002). Although, rainfall and groundwater recharge in desert regions is small, the total amount of recharge could be considerable, given that potential recharge areas are large. In order to provide a sound basis for research on recharge processes in desert regions, spatial estimates of rainfall fields and rainfall intensity are required (de Vries and Simmers 2002). However, reliable rainfall data sources are often only available as point data, such as rainfall data from meteorological stations. The density of meteorological stations in desert regions is limited so that spatial interpolation of rainfall data does not fully represent the high spatial and temporal variability of rainfall in the Arabian Peninsula (Barth and Steinkohl 2004), such meteorological stations are usually present on the airports. Friesen et al. (2010) used an approach by which they identified rainfall events and estimated their spatial extent using remote sensing data. Satellite-derived vegetation estimates were used to validate rainfall occurrences, and to include or exclude water losses through plants (transpiration) in addition to losses from soil (evaporation). Rainfall fields, based on 11 years of TRMM 3B42 data have been extracted for the whole Arabian Peninsula, this was used to illustrate the extraction of regional rainfall statistics, as well as the validation methodology, as an example, a subset of the data representing the Rub Al Khali desert region was successfully used to achieve the objective.

A GIS-based water balance model has been developed (Sherif 2009) for water availability and demands in UAE (quantitatively and qualitatively). The function of water availability includes three main components: groundwater, desalinated water and the TWW. In addition, surface water potentiality has also been addressed. The study aimed at assessing the current status of supply sources and demand centers, in terms of their current capacity to supply water (sources) and their current requirements for water (demand centers) and providing a planning tool to allow rapid judgment of the future impacts of changes in supplies and demands on the supply/demand balance (Decision Support Model). The demand is categorized into domestic demand, agriculture demand, amenity demand, and forestry demand. Calculation of each demand is made by using an appropriate mathematical model. Domestic demand is determined from population and per-capita water consumption, both specified for individual years of the simulation period adopted in the SDM. The water requirements for agriculture are calculated from net crop consumptive use of individual crops, percentages of the crops within crop mix and irrigation efficiency. The components of the water resources supply for this modeling purpose included; surface water supply mainly through rainfall water harvested with dams or other infrastructure, the groundwater supply, desalinated water supply and the TWW supply. The newly developed water budget module was rigorously tested and water budgets calculated for Abu Dhabi for the years 1996 and 2007.

Other researchers (Nouh 1990; Sorman and Abdulrazzaq 1993a, b), in their studies in Saudi Arabia, have used an advanced metric model (GIUH model) and found to be highly affected by some parameters that needed to be directly measured from the field, which makes it difficult to use in catchments with limited data. The KINEROS, a physically-based, event oriented model can be considered as one of the most important models developed to consider arid conditions. In the Sultanate of Oman, Wheater (1981) has used the model successfully, after a heavy calibration, and the obtained results were in a good agreement with the calculated peak discharge and the estimated time to peak. However, Michaud and Sorooshian (1994) found that the model cannot cope with medium to large catchment areas. McIntyre et al. (2007) applied regression models to predict flood peaks and volumes in Wadi Ahin, using essentially 25 events but splitting some of the longer events, which would not run in KINEROS2 due to numerical problems, to give a total of 34 events in their regression. They found that, by linear regression of flow peak and volume against gauge rainfall, 16 of 34 observed flow peaks were predicted to within 30 % and 11 of 34 observed flow volumes were predicted to within 30 %. Al Qurashi (2010) discussed the effects of the availability and quality of the data used on the performance of the models selected and on the results obtained using KINEROS2. The model was applied to hourly data from two different type runoff events in Wadi Ahin (734 km^2 catchment area) in the Sultanate of Oman. The results showed that the performance of the model was highly dependent on the availability of the input data and its quality. The models' outputs demonstrate the difficulty of obtaining highly reliable results and raises the question as to the extent such results can be depended upon as a basis for important planning and management decisions. Some recommendations were provided on how models' performances and their results can be improved.

In Oman, the major agricultural region is Al-Batinah, located along the coast beginning north of the capital Muscat. The area under cultivation reaches approximately 31,100 ha representing more than 50 % of the country total. The concern in Al-Batinah is that increased competition for water is resulting in salt water intrusion into the coastal aquifer (Kacimov et al. 2009). In addition, as urban populations expand in the Al-Batinah region, there is increasing pressure for producers to move to areas where land is less expensive. As a result, political and economic pressures are being exerted on agricultural water users to increase irrigation efficiency. One way to achieve this objective is an improvement of water productivity which needs a good quantitative understanding of the relationship between irrigation practices and grain yield, i.e. crop water production function (CWPF). With this knowledge, the value of each unit of water applied to a field can be estimated and compared with alternative uses within and beyond the agricultural sector. In arid climates, irrigation methods which reduce watering volumes and, at the same time, maximize WUE were developed to coexist with the increasing reduction of water resources and the increase in irrigation costs (Jones 2004).

4.4.2.2 Water Salinity Management Models

Efficient management of salinity problems in irrigated areas necessarily requires detailed understanding of the processes and factors controlling the movement of water and salts to the groundwater, with special emphasis on the root zone where irrigation water is used to offset crop water requirements. In the GCC region limited work has been done in the development of models based on local field conditions. In the Batinah region of Oman, Al Ajmi et al. (2002) used historical data set coupled with limited direct field measurements to build a soil salinization model based on the sound physical concepts, rather than developing a statistical model, and which is capable of being used with limited data available. The objective of this study was to develop a model to explain the variation between farms in soil-water salinities, and thus this has provided a tool to optimize management practices (amount and frequency of irrigation, and the sizes of irrigation basins used in the case of the tree crops) to control soil salinization. Within the farms studied, there was wide variation in each of these practices, which is likely to be an important factor influencing the observed variation in soil-water salinity. A major challenge of this study was the limited, and raw and even absence of data about the factors (amount and frequency of irrigation, as long-term monitoring was impractical). The objective was to study small farms (rather than research stations) without information on irrigation scheduling and irrigation water quality. Direct measurements were limited to data from soil and irrigation water salinity surveys carried out in the South Batinah Integrated Study (MAF 1993) and the other measurements from this study. A model termed IMAGE (irrigation management model) was developed that could make remarkably good predictions of soil-water salinity, with 75 % of locations being predicted to within 2 dS m^{-1}, in the context of absolute values ranging from 1.93 to 51.1 dS m^{-1}. Poorly predicted sites were those with large data uncertainties, or factors such as highly permeable soils where there was large within-basin heterogeneity in infiltration and drainage. Sensitivity analyses showed there to be considerable scope for controlling soil-water salinity to within acceptable limits through optimizing irrigation scheduling and the size of irrigation basins relative to the size of the tree canopy.

4.4.3 Use of Treated Waste Water (TWW) in Irrigation

The TWW is an assured resource and the only one that is also guaranteed to increase in response to population growth. The limited availability of traditional water resources necessitates the use of non-traditional resources (TWW) in agriculture. The current primary use of TWW in GCC countries is for municipal landscaping while a significant volume is lost to the sea even after it is treated to the secondary level (World Bank 2005). Currently, about 60 % of the sewage effluent in GCC countries is treated but its use in agriculture accounts for only 2 % of the total amount of water used in crop production (World Bank 2005). Wastewater reuse in agriculture is an important approach to help overcome the water scarcity problems of the region. Wastewater reuse in the region could contribute significantly to solving the problems of quality and quantity of water. However, it depend on how it is used?. In terms of costs, it is the most feasible option to augment the water resources in the region. As a result, wastewater reuse will have a major impact on the agricultural economy, as well as on the well-being and the health of the society (Al Salem and Abouzaid 2006).

Wastewater is increasingly used in agriculture activities both in developing and developed countries. This use is due to the increasing water scarcity and stress, the population increase and related increased demand for food and fiber and the degradation of freshwater resources (WHO 2006). Most population growth is expected to occur in the urban and peri urban areas in the developing countries (United Nations (UN) 2002). Population growth increases both the demand for fresh water and the amount of wastes that are discharged into the environment, thus leading to more pollution of clean water sources. Wastewater, upon proper treatments, is often a reliable year-round source of water; it also contains some of the essential nutrients (NP) required for plant growth. Globally (Schipper et al. 2006) at least two million ha are presently irrigated with untreated, partially treated, diluted, or TSE. Taha (2001) studied the impacts of irrigation by TSE (for 10, 20 and 30 years) on the soil properties of the Green Belt. The results of the study indicated that no significant changes in soil mineral content (particularly phosphorus) and particle size distribution were induced. Taha et al. (2002) studied the impacts of irrigation with TSE on forage production and concluded that the practice improved plant growth and yield significantly. Taha (2001) proposed some precautious measures in association with this practice such as to avoid over-irrigation since excessive water tends to harbor slime and algae that may suffocate the plant and decrease downward movement of water, plants are to be planted at the shoulders of the ridges as flat beds could contribute to death by water logging and farmers should be aware of the hazards of working with TSE and adopt safety measures. They should wear special boots covering the legs, avoid direct contact with TSE and wash and clean themselves when they get out of the fields irrigated with TSE.

4.4.4 Utilizing Modern Technologies of Soilless and Hydroponic Culture

High on-farm WUE depends on adequate and timely supplies of water (Incrocci et al. 2003). Soilless culture has been used as a tool to save water and fertilizer in comparison to cultivation in the soil as open field agriculture (De Rijk and Schrevens 1998). Most soils in the GCC region suffer from salinity to various degrees, low organic matter due to scarce biodiversity, low inherent soil fertility (low clay content) and low nutrient and water retention capacity and high infiltration rate leads to increased deep percolation. The soilless technique is based on cultivating crops outside the soil environment. Since soils supply plants with the required nutrients, these nutrients are dissolved in irrigation water under soilless cultivation method. Various media have been tested as an alternative to soils to anchor plants and as moisture reservoirs. The zeolite, rock wool, pumice and perlite are few examples. However, the unique adsorption, cation exchange capacity (CEC), water sorption capacities and dehydration-rehydration properties of natural zeolites make them most suitable for use in soilless cultivation (Pisarovic et al. 2003). To avoid wastages through deep percolation, plants are grown within a confined root zone. Drainage water can then be re-circulated through the system to attain maximum efficiency.

Hydroponics is a subset of soilless culture and is a method of growing plants using mineral nutrient solutions, in water, without any growing substrate. Roots of the plants are soaked into the water in which nutrient solution is dissolved. This method is highly potential for producing sprouts of various fodder crops from seeds in a short period of time of less than 10 days. Unlike field production system that use run-to-waste irrigation practices, the hydroponic fodder system uses recirculation system, thus reducing the waste water. It has been reported that hydroponic fodder production requires about 2–3 % of the water used under field conditions to produce the same amount of fodder (Al-Karaki and Al-Momani 2011). Fodder produced hydroponically is of a short growth period (7–10 days) and does not require high-quality arable land, but only a small piece of land for production to take place (Cuddeford 1989; Mooney 2005). The fodder is of a high feed quality, rich with proteins, fiber, vitamins, and minerals (Bhise et al. 1988; Lorenz 1980). All these special features of hydroponic system, in addition to others make it one of the most important agricultural techniques currently in use for green forage production in GCC region and many other countries. Hydroponic technique can be used for green fodder production of many forage crops in a hygienic environment free of chemicals like insecticides, herbicides, fungicides, and artificial growth promoters (Jensen and Malter 1995; Al-Karaki and Al-Momani 2011). It is a well-known technique for high fodder yield, year round production and least water consumption (Al-Karaki 2011).

4.5 Sultan Qaboos Center for Modern and Soilless Culture-SQCMSC

The SQCMSC center was established in 2001 at the Arabian Gulf University – a GCC regional university in Bahrain (Fig. 4.2a, b). The center was started to promote research in modern agricultural techniques with potential to produce more crops with less water and to adopt these techniques for the region. The ultimate aim was to utilize innovative technologies to achieve agricultural sustainability and produce crops at competitive quality and price. Soilless technique was considered as one of the most encouraging methods of crop production in the region especially for high market value crops which are mostly imported. Locally produced crops have the advantages over the imported ones in many aspects such as savings in the cost of production due to cheap labor market, savings in overseas transportation cost, providing the market with more flourishing flowers and fresh products and creating employment opportunities for local people. The investments in this type of business have the opportunities in attracting young graduates from higher secondary schools (both males and females) a category of labor force which comprises the majority of the jobless citizens in the region, especially in the Sultanate of Oman.

The center was established with a vision that the region should have a sustainable agricultural sector and a tributary of the national economy that contributes to food security. The mission was to contribute to the efforts by the GCC countries to achieve sustainable agricultural development in the light of limited natural resources, shifting the agricultural sector in the region to an ever-growing sector and a tributary of the economy that create employment opportunities for youth and recruiting methods and techniques of modern agriculture such as soilless agriculture in order to maximize agricultural production and preserve water resources. The work plan for the center was designed to attain several goals, (1) to create opportunities for researchers and higher education students to specialize in agricultural techniques and hydroponics systems, (2) to conduct practical research in using advanced agricultural techniques especially hydroponics, (3) to build the capacity of the relevant people through seminars and training workshops specialized in advanced agricultural techniques, (4) to present experiences, give consultations and conduct studies in the field of modern agricultural techniques and, (5) to make contacts and establish connections between the Arabian Gulf University and other universities, relevant ministries and institutions in the GCC region.

Fig. 4.2 Plant production in SQCMSC. (**a**) Green houses at SQCMSC. (**b**) Green house experiments. (**c**) Flowers production. (**d**) Hydroponic experiments. (**e**) Forage production using tap and TWW

4.5.1 Major Research Interests and Achievements at SQCMSC

The SQCMSC center conducted research studies over three phases. Because of the lack of baseline data in the region on crop performance under soilless and hydroponic, most of the research work at the initial stage was focused on screening the types of crops which are likely to be potential for the region from physical and economical perspectives. A large number of vegetable crops such as cherry tomatoes, cucumbers, okras, iceberg lettuce, egg plants, capsicums and flowers like gerberas, eustomas, chrysanthemums and many others were tested and "fact sheets" about their production in terms of inputs and outputs were produced. The second phase of the research plan aimed at conducting more in-depth research studies about the various aspects related to the adaptation of these modern techniques of production in the region. A number of studies were conducted and published in highly recognized peer reviewed refereed journals and, also the results were presented in international conferences (Fig. 4.2c).

It should be kept in mind while developing solution for hydroponic system, that, the salinity of water used to prepare nutrient solution should be low enough to be able to obtain a final solution with a reasonable salinity which is tolerable by the crops grown, usually below the salinity threshold value of crop in use (Benton 1930). The center conducted studies to determine the salinity thresholds for some of the potential crops grown under this kind of growing systems. For example, Al-Karaki et al. (2009) examined the response of three sweet pepper cultivars to different levels of irrigation water salinity under soilless recirculating conditions using zeolite as growing media. The sweet pepper proved to be sensitive to the increasing salinity. However, genotype variations for fruit yield and yield components have been noticed.

Another concern when using soilless techniques is the relatively high cost of imported substrates (growing media). Many such media are used nowadays in soilless production systems. The cost of these media is an important component in total production cost. The main criteria for selection of a particular substrate are the substrate cost and its favorable chemical and physical characteristics being kept for long time (Al-Ajmi et al. 2009a). This was another research aspect which the center has invested in. For instance, Al-Ajmi et al. (2009a) compared the efficiency of three different substrates namely zeolite, perlite and sand and three combinations of them (mixed substrates) on cherry tomatoes in terms of performance and relative cost. Differences in fruit yield and quality were observed among the substrates used with the highest performance obtained by zeolite alone probably due to its high water holding capacity and cation-exchange-capacity (Pisarovic et al. 2003). However, the results indicated that addition of sand to zeolite at (1:1) volume basis ratio resulted in no significant difference to zeolite alone. This finding is of much interest and cost effective to the GCC region where zeolite is less available whereas sand is abundantly available.

The third field of research was focused on the use of hydroponic growing systems to produce fodder crops (Fig. 4.2d). Fodder crops, like Alfalfa and Rhodes grass, though highly thirst to water, are extensively cultivated in the region. Reducing agricultural water use while maintaining or improving economic productivity of the agricultural sector is a major challenge in this region. Over recent years, severe shortages in food supplies for livestock have been experienced in GCC countries mainly due to repeated droughts as well as shortages of water for irrigation. Many projects to produce forages have been established during the last two decades to cover some green and dry forage needs in these countries. However, scarcity of adequate fresh water supplies might pose challenges for sustainability of the field projects especially with utilizing ground water for irrigation, which is consumed in large amounts as these countries are characterized with very high rates of evapotranspiration and soils of low capacity to retain water (Al-Karaki and Al-Hashimi 2012). Therefore, methods and technologies that can contribute to improved water use efficiency and productivity merit closer consideration like hydroponic technique. A study by Al Ajmi et al. (2009b) investigated yield and WUE of the hydroponically produced barley sprouts using tertiary treated sewage effluent (TTSE). Seeds were sown in stacked trays in a temperature controlled room. Trays were irrigated daily with either tap water, or tap water mixed with TTSE at 20, 40, 60, 80 % and with TTSE only. Results indicated that germination percent and yield of barley increased as the concentration of TTSE in irrigation water increased, however, the increase in WUE was non-significant. Proximate chemical analyses indicated that there was no significant effect of TSE on moisture, crude fiber, acid detergent fiber, neutral detergent fiber, or fat (ether extract) of the barley fodder. It was concluded that barley produced by TTSE maintained all its fodder quality and that it can be produced commercially for feeding livestock. This study was conducted with joint efforts from SQCMSC and the SQU in Oman.

The study further investigated the possible effects of TTSE (Fig. 4.2e) on forage biomass in terms of the accumulation of heavy metals (mainly Pb, Cd and Ni) and other nutrient elements (like N, P, K, Ca and Fe) since all these elements were present in much higher concentrations in TTSE compared to tap water (Al-Ajmi et al. 2009b). Though heavy metals concentrations in barley biomass increased with an increase in the concentration of TTSE (Al-Ajmi et al. 2009c), they were within the limits set by the Commission of the European Union and WHO (EC 2003). The concentrations of N, P, K, Ca and Fe in barley biomass were also within the normal limits. The study revealed that fodder barley grown hydroponically could be irrigated safely with TTSE, as a useful alternative disposal method of wastewater without the risk of accumulation of heavy metals in the soil.

The third phase of research commenced late 2011. It focused on testing the economic feasibility of crop production under soilless and hydroponic systems. To attain this goal, Arabian Gulf University established six new green houses and added to the existing four to provide the essential basis for determining the influence of economy of scale parameter on the overall feasibility of crop production. Data have been collected on all inputs and outputs to be able to accomplish these calculations.

4.5.2 Business Incubators-a Step Forward Towards the Dissemination of Hydroponic Techniques Among Young Investors in the GCC Countries

Investment in hydroponic based crop production business can open job opportunities for young and fresh graduates in the GCC countries, who constitute a major proportion of jobless. Such opportunities will enhance locally produced agricultural commodities leading to bridge gap between imported and local agriculture commodities and hence this will be a way forward to achieve food security. To be successful in such enterprise an initial training must be given to the new investors. This necessitates interventions by governments for support and encouragement through business incubators mechanism in which the government establishes the infrastructure required for hydroponic growing systems and attracts a potential group of interested and serious young individuals and provide them with the necessary training, guarantees the market for their products and work with them in a joint venture business through legal contracts. As these workers obtain the necessary skills in managing the business, the government reduces its intervention and shares to allow them to gradually take over the job and eventually own the business within a certain timeframe. The investors pay back the cost for the infrastructure to the government in affordable installments within an agreed framework. The government could then start the second and the following stages in this business to expand the technology to as many people as possible.

4.6 Conclusions

Agricultural sustainability in the GCC region is threatened by many constraints including but not limited to water scarcity, harsh climatic conditions, insufficient arable lands, poor irrigation management and climate change impact. However, opportunities exist for reversing the situation before it is irreversible if strong policies for better utilization and management of water are established and backed up with effective legislations and procedures for implementation. Management of water resources in the region has so far emphasized the supply side by expanding water supply facilities and structures such as desalination plants and new recharge dams. Experience from the past shows that this is not going to be sustainable from economic and environmental perspectives. Management of water resources on the demand side has better opportunities to sustain the various sectors involved such as mainly agriculture. Shifting the current traditional irrigation practices to more efficient and climate smart irrigation (CSA) methods can save appreciable amount of water wastages by deep percolation and evaporation. Water and salinity models can be used where appropriate to understand and future forecasting of water recharge, management, and planning research purposes. Models can also be an efficient tool for managing salinity problems in irrigated area. The limited availability of conventional water resources necessitates the use of non-conventional water resources in agriculture among which treated wastewater (TWW) is gaining importance due to many reasons. The current primary use of TWW in GCC countries is for municipal landscaping while a significant volume is lost to the sea even after it is treated to the secondary or tertiary levels. Since high on-farm WUE depends on adequate and timely supplies of water, soilless culture can be used as a tool to save water and fertilizer in comparison to open field cultivation. Sultan Qaboos Center for Modern and Soilless Culture SQCMSC was established at the Arabian Gulf University – a GCC regional university in Bahrain- to promote research in modern agricultural techniques with potential to produce more crops with less water and to adopt these techniques for the region. The ultimate aim was to utilize innovate technologies to achieve agricultural sustainability and produce crops at competitive quality and price. Business incubators sponsored by GCC governments for hydroponically produced crop commodities are considered as a potential approach to disseminate this technology among young graduates, create job opportunities and achieve better food security for the region.

References

Al-Ajmi A, Nortcliff S, Simmonds LP (2002) IMAGE; a physically based one dimensional irrigation management model for soil salinity control. Soil Use Manag 18:184–190

Al-Ajmi A, Al-Karaki G, Othman Y (2009a) Effect of different substrates on fruit yield and quality of cherry tomato grown in a closed soilless system. Acta Hortic 807:491–494

Al-Ajmi A, Saleh A, Kadim I, Osman Y (2009b) Chemical constituents and heavy metals contents of barley fodder produced under hydroponic system in GCC countries using tertiary treated sewage effluents. J Phytol 1(6):374–380

Al-Ajmi A, Saleh A, Kadim I, Osman Y (2009c) Yield and water use efficiency of barley fodder produced under hydroponic system in GCC countries using tertiary treated sewage effluents. J Phytol 1(5):342–348

Al-Karaki GN (2011) Utilization of treated wastewater for green forage production in a hydroponic system. Emir J Food Agric 23:80–94

Al-Karaki GN, Al-Hashimi M (2012) Green fodder production and water use efficiency of some forage crops under hydroponic conditions. ISRN Agron. Article ID 924672, 5 pp. doi:10.5402/2012/924672

Al-Karaki GN, Al-Momani N (2011) Evaluation of some barley cultivars for green fodder production and water use efficiency under hydroponic conditions. Jordan J Agric Sci 7(3):448–456

Al-Karaki G, Al-Ajmi A, Othman Y (2009) Response of soilless grown sweet pepper cultivars to salinity. In: Tuzel Y et al (eds) Proceedings of the international seminar on protected cultivation. Mild winter climate. ACTA Hortic 807; 1:227–231. ISHS

Al-Omran AM, Sheta AS, Falatah AM, Al-Harbi AR (2005) Effect of drip irrigation on squash (Cucurbita pepo) yield and water-use efficiency in sandy calcareous soils amended with clay deposits. Agric Water Manag 73(1):43–55

Al-Omran AM, Al-Harbi AL, Wahb-Allah MA, Nadeem M, Al-Eter A (2010) Impact of irrigation water quality, irrigation systems, irrigation rates and soil amendments on tomato production in sandy calcareous soil. Turk J Agric For 34:59–73

Al-Qurashi AM (2010) Rainfall-runoff modeling performance in arid areas under scarce hydrological data. Paper presented at the WSTA 9th Gulf water conference, Muscat, Sultanate of Oman, 22–25 Mar 2010

Al-Rashed MF, Sherif MM (2000) Water resources in the GCC countries: an overview. Water Res Manag 14:59–75

Al Salem SS, Abouzaid H (2006) Wastewater reuse for agriculture: regional health perspective. East Mediterr Health J 12(3–4):446–458

Alsharhan AS, Rizk ZA, Nairn AEM, Bakhit DW, Alhajari SA (2001) Hydrogeology of an arid region: the Arabian Gulf and adjoining areas. Elsevier Science, Amsterdam, 331 pp

ASAE Standards (1999) S 526.1, soil and water terminology. American Society of Agricultural Engineers Standards, St. Joseph

Ayars JE, Phene CJ, Hutmacher RB, Davis KR, Schoneman RA, Vail SS, Mead RM (1999) Subsurface drip irrigation of row crops: a review of 15 years of research at the Water Management Research Laboratory. Agric Water Manag 42:1–27

Barth HJ, Steinkohl F (2004) Origin of winter precipitation in the central coastal lowlands of Saudi Arabia. J Arid Environ 57(1):101–115

Bazza M (2005) Policies for water management and food security under water-scarcity conditions: the case of GCC countries. Paper presented at the 7th Gulf water conference, Kuwait, 19–23 Nov 2005

Benton J (1930) Hydroponics; a practical guide for the soilless grower, 2nd edn. CRC Press, Boca Raton, p 455

Bhise V, Chavan J, Kadam S (1988) Effects of malting on proximate composition and in vitro protein and starch digestibilities of grain sorghum. J Food Sci Technol 25(6):327–329

Camp CR (1998) Subsurface drip irrigation: a review. Trans ASAE 41(5):1353–1367

Charlesworth PB, Muirhead WA (2003) Crop establishment using subsurface drip irrigation: a comparison of point and area sources. Irrig Sci 22(2):195–200

Condon AG, Richards RA, Rebetzke GJ, Farquhar GD (2002) Improving intrinsic water-use efficiency and crop yield. Crop Sci 42:122–131

Cuddeford D (1989) Hydroponic grass. Practice 11(5):211–214

De Rijk G, Schrevens F (1998) Distribution of nutrients and water in rockwool slabs. Sci Hortic 72:277–285

de Vries JJ, Simmers I (2002) Groundwater recharge: an overview of processes and challenges. Hydrogeol J 10(1):5–17

Demir AO, Goksoy AT, Bugkcangaz H, Turan ZM, Koksal ES (2006) Deficit irrigation of sunflower (Heliauthus annuus L.) in a sub-humid climate. Irrig Sci 24(4):279–289

EAD (2009a) Soil survey of Abu Dhabi Emirate, vol I, Extensive survey. Environment Agency-Abu Dhabi, Abu Dhabi

EAD (2009b) Soil survey of Abu Dhabi Emirate, vol III, Intensive survey. Environment Agency-Abu Dhabi, Abu Dhabi

EC (2003) Risk assessment report on cadmium metal and cadmium oxide: final draft version. European Commission.

El-Shafei YZ (1989) Response of tomatoes to trickle irrigation on sandy soil in an arid zone in Libya. Arid Soil Res Rehabil 3:477–486

ESCWA (United Nations Economic and Social Commission for Western Asia) (2009) ESCWA water development report 3: role of desalination in addressing water scarcity. E/ESCWA/SDPD/2009/4. ESCWA, New York

Friesen J, Hildebrandt A, Kalbus E, Attinger S (2010) Rainfall fields in the interior desert regions of the Arabian Peninsula. Paper presented at the WSTA 9th Gulf water conference, Muscat, Sultanate of Oman, 22–25 Mar 2010

Gajre PR, Gill KS, Chaudhary MR (1997) Irrigation of sunflower (Heliauthus annuus L.) in relation to tillage and mulching. Agric Water Manag 34(2):149–160

GCC General Secretariat official website (2012) http://www.gcc-sg.org/eng/. Accessed 17 Apr 2012

Hansen JE, Lacis AA, Lee P, Wang WC (1980) Climatic effects of atmospheric aerosols. Ann N Y Acad Sci 338:575–587
Harahsheh H, Mashroom M, Marzouqi Y, Al Khatib E, Rao BRM, Fyzee MA (2013) Chapter 5: Soil thematic map and land capability classification of Dubai Emirate. In: Shahid SA, Taha FK, Abdelfattah MA (eds) Developments in soil classification, land use planning and policy implications-innovative thinking of soil inventory for land use planning and management of land resources. Springer, Dordrecht, pp 133–146
Hutmacher PB, Mead RM, Shouse P (1996) Subsurface drip: improving alfalfa irrigation in the west. Irrig J 45:48–52
Incrocci L, Pardossi A, Campiotti CA, Balducchi R, Guinchi L (2003) Energy, water and fertilizer requirements of a closed loop soilless culture of green house cherry tomato in Cicily. Acta Hortic 614:189–192
James LG (1988) Principles of farm irrigation systems design. Wiley, New York, p 543
Jensen ME (ed) (1981) Design and operation of farm irrigation systems, ASAE monograph, No. 3. ASAE, St. Joseph, pp 829
Jensen MH, Malter AJ (1995) Protected agriculture: a global review, vol 253. World Bank Publications, Washington, DC
Jin MG, Zhang RQ, Sun L, Gao YF (1999) Temporal and spatial soil water management: a case study in Heilonggang region. Agric Water Manag 42(2):173–187
Jones HG (2004) Irrigation scheduling: advantages and pitfalls of plant-based methods. J Exp Bot 55(407):2427–2436
Kacimov AR, Sherif MM, Perret JS (2009) Control of sea-water intrusion by salt-water pumping: coast of Oman. Hydrogeol J 17(3):541–558
Keller J, Bliesner RD (1990) Sprinkler and trickle irrigation. Van Nostrand Reinhold, New York, pp 652
Kiang TT (2008) Singapore's experience in water demand management. http://www.iwra.org/congress/2008/resource/authors/abs461_article.pdf
KISR (1999) Soil survey for the State of Kuwait, 5 vols. Kuwait Institute for Scientific Research, Kuwait
Lamm FR, Trooien TP (2003) Subsurface drip irrigation for corn production: a review of 10 years of research in Kansas. Irrig Sci 22:195–200
Lorenz K (1980) Cereal sprouts: composition, nutritive value, food applications. Crit Rev Food Sci Nutr 13(4):353–385
MAF (1990) General soil map of the Sultanate of Oman. Ministry of Agriculture and Fisheries, Directorate General of Agricultural Research, Muscat
MAF (1993) South Batinah integrated study. Directorate General of Agricultural Research, Ministry of Agriculture and Fisheries, Oman
MAW-Ministry of Agriculture and Water (1985) General soil map of the Kingdom of Saudi Arabia. Ministry of Agriculture and Water, Riyadh
McClatchy (2007) GCC: oil revenue bonanza has boosted government spending. Middle East and North Africa business report, Tribune Information Services. Retrieved 10 April 2014 from http://www.highbeam.com/doc/1G1-165990278.html
McIntyre N, Al-Qurashi A, Wheater HS (2007) Analysis of rainfall-runoff events from an arid catchment in Oman. Hydrol Sci J 52(6):1103–1118
Michaud J, Sorooshian S (1994) Comparison of simple versus complex distributed runoff models on midsized semiarid watershed. Water Res 30:593–605
Miller RJ, Replston DE, Rauschkolb RS, Wolfe DW (1981) Labeled nitrogen uptake by drip irrigated tomatoes. Agron J 73(2):265–270
Mooney J (2005) Growing cattle feed hydroponically. 2002 scholarship report, Australian Nutffield Farming Scholars Association, 30 p
Musick JT, Jones OR, Stemart BA, Dusek DA (1994) Water yield relationships for irrigated and dry land wheat in the US southern plains. Agron J 86(6):980–986
Nouh M (1990) Flood hydrograph estimation from arid catchment morphology. Hydrol Process 4:103–120

Omar SAS, Shahid SA (2013) Chapter 3: Reconnaissance soil survey for the State of Kuwait. In: Shahid SA, Taha FK, Abdelfattah MA (eds) Developments in soil classification, land use planning and policy implications-innovative thinking of soil inventory for land use planning and management of land resources. Springer, Dordrecht, pp 85–107

Phene CJ, McCormick RL, Davis KR, Oierro JA (1989) Lysimeter system for precise evapotranspiration measurements and irrigation control. Trans ASAE 32:477–484

Pisarovic A, Filipan T, Tisma S (2003) Application of zeolite based special substrates in agriculture: ecological and economical justification. Period Boil 105:287–293

Sammis TW (1980) Comparison of sprinkler, trickle, subsurface and furrow irrigation methods for row crops. Agron J 72(5):701–704

Scanlon BR, Healy RW, Cook PG (2002) Choosing appropriate techniques for quantifying groundwater recharge. Hydrogeol J 10(1):18–39

Scheibert C, Stietiya MH, Sommer J, Abdalla OES, Schramm H, Memah Al M (2005) The atlas of soils for the State of Qatar. Ministry of Municipal Affairs and Agriculture, Doha

Schipper L, Moore M, Molden D (2006) Research priorities. CGIAR Challenge Program on Water and Food (CPWF) and Comprehensive Assessment of Water Management in Agriculture (CA)

Schwankl LJ, Grattan SR, Miyao EM (1990) Drip irrigation burial depth and seed planting depth effects on tomato germination. In: Proceedings of the third national irrigation symposium. ASAE, St. Joseph, pp 682–687

Shahid SA, Taha FK, Abdelfattah MA (eds) (2013a) Developments in soil classification, land use planning and policy implications-innovative thinking of soil inventory for land use planning and management of land resources. Springer, Dordrecht, 858 pp

Shahid SA, Abdelfattah MA, Othman Y, Kumar A, Taha FK, Kelley JA, Wilson MA (2013b) Chapter 1: Innovative thinking for sustainable use of terrestrial resources in Abu Dhabi Emirate through scientific soil inventory and policy development. In: Shahid SA, Taha FK, Abdelfattah MA (eds) Developments in soil classification, land use planning and policy implications-innovative thinking of soil inventory for land use planning and management of land resources. Springer, Dordrecht, pp 3–49

Sherif MM (2009) Development of a water balance model as precursor for decision support system for water resources management. Phase two: water balance model operation manual. Schlumberger Water Services, Abu Dhabi

Simsek M, Tonkaz T, Kacira M, Comlekcioglu N, Dogan Z (2005) The effects of different irrigation regimes on cucumber (cucumbis sativus L.) yield and yield characteristics under open field conditions. Agric Water Manag 73(3):173–191

Sorman AU, Abdulrazzaq J (1993a) Infiltration-recharge through wadi beds in arid region. Hydrol Sci J 38:173–186

Sorman AU, Abdulrazzaq J (1993b) Flood hydrographs estimation for ungauged wadis in Saudi Arabia. Water Res Plan Manag 119–1:45–63

Sun H, Liu C, Zhang Y, Shen Y, Zhang Y (2006) Effects of irrigation on water balance, yield and WUE of winter wheat in the North China plain. Agric Water Manag 85(1–2):211–218

Taha IM (2001) Impacts of treated sewage water on soil characteristics, forage production and the surrounding inhabitants. PhD thesis, Faculty of Agriculture, University of Khartoum, Sudan

Taha IM, Hamza NB, Malik MN (2002) Utilization of treated sewage water forage production. In: Proceedings of the international symposium on environment pollution control and waste management, Tunis, 7–10 Jan 2002, pp 560–572

Tiwari KN, Mal PK, Singh RM, Chattopadhyay A (1998) Feasibility of drip irrigation under different soil covers in tomato. J Agric Eng 35(2):41–49

United Nations (UN) (2002) Arab human resource development. UN Publications, New York

Walker WR, Skogerboe GV (1987) Surface irrigation. Prentice Hall, Englewood Cliffs, 386 pp

Wallach R (1990) Effective irrigation uniformity as related to root zone depth. Irrig Sci 1:15–21

Wheater HS (1981) Northern Oman flood study. Imperial College, London, 270 pp

WHO (2006) Guidelines for the safe use of wastewater, excreta and grey water. WHO Press, France

World Bank (2005) Water sector assessment report on the country of the cooperation of the Gulf. World Bank documents. Report no. 32539 – MNA, 83 pp

World Bank (2012) MENA development report. Overview: renewable energy desalination: an emerging solution to close the water gap in the Middle East and North Africa, Washington, DC, pp 1–18

WWAP (World Water Assessment Programme) (2009) World water development report 3: water in a changing world. UNESCO/Earthscan, Paris/London

WWAP (World Water Assessment Programme) (2012) The United Nations world water development report 4: managing water under uncertainty and risk. UNESCO, Paris

Zubari WK (1997) Towards the establishment of a total water cycle management and re-use program in the GCC countries. Paper presented at the 7th regional meeting of Arab IHP committees, Rabat, Morocco, 8–12 Sept 1997

Chapter 5
The Potential of Transforming Salalah into Oman's Vegetables Basket

Sanmugam A. Prathapar, Mumtaz Khan, and Msafiri Daudi Mbaga

Abstract Agriculture in the Sultanate of Oman is mostly small scale and is a part of the traditional way of life. The majority of the population benefit from agriculture, however little. The 67 % of the population was in households that had at least one crop or livestock holding where the output contributed to consumption or income. Since the year 2000, the Government spent Rial Omani (RO) 20.1 million on agriculture and fishery development, and another RO 39.4 million on water resources development. Furthermore, the government encourages farming by offering land, machinery, and extension services. However, during the period 2000 till 2007, crop production has in fact gone down. In other words, despite being a capital rich country, substantial investment in agriculture, it is increasingly becoming a food insecure country. An in-depth analysis of Oman's agricultural sub-sectors shows that, household sub-sector contributed 27 % of the total value. Primary crop production in Oman in 2005/2007 was 486.872 metric tons of which contribution of fruits and vegetables were 353,072 metric tons and 102,606 respectively. In comparison, only 26,206 metric tons of cereals were produced. The value of production of cereals and vegetables were 7.8 and 17.6 million RO respectively. This comparison confirms that Omanis prefer producing high value vegetables to cereal crops. In addition to vegetables produced locally, Oman imported 148,345 metric tons during the same period. Therefore, it is interesting to explore, if vegetable

S.A. Prathapar (✉)
Theme Leader, International Water Management Institute, New Delhi, India
e-mail: S.Prathapar@cgiar.org

M. Khan
Department of Crop Sciences, College of Agricultural and Marine Sciences,
Sultan Qaboos University, Muscat, Oman
e-mail: mumtaz@squ.edu.om

M.D. Mbaga
Department of Natural Resource Economics, College of Agricultural and Marine Sciences,
Sultan Qaboos University, Muscat, Oman
e-mail: msafiri@squ.edu.om

S.A. Shahid and M. Ahmed (eds.), *Environmental Cost and Face of Agriculture in the Gulf Cooperation Council Countries: Fostering Agriculture in the Context of Climate Change*,
DOI 10.1007/978-3-319-05768-2_5,

production in Oman can be further increased, resulting in increased income and near self-sufficiency in vegetables. If Oman chooses to increase vegetable production, then it has to come from a major shift in its current land and water use practices, because almost all of its cultivable lands and available freshwater are fully utilized at present. In this chapter we explored if the Salalah region of Oman could be transformed into Oman's vegetable basket, leading to self-sufficiency in its vegetable needs.

Keywords Food import • Salalah • Vegetables basket • Oman • Water availability

5.1 Introduction

The World Bank in association with the Food and Agriculture Organization (FAO) of the United Nations (UN) and the International Fund for Agricultural Development (IFAD) has comprehensively deliberated on the issue of improving food security in Arab countries given the recent experience of the food price surge (World Bank 2009). It identifies Arab countries as the most vulnerable to food insecurity, given rapidly increasing demand for food, limited capacity for domestic food production and thus heavy dependence on world food markets. The study projects that dependence of food imports in the Arab region would increase by 64 %. A three prong strategy to secure food security has been proposed. These include, (i) strengthening food safety nets such as through improved family planning and education, (ii) reducing vulnerability to international food market vagaries through adoption of improved supply chain management and use of financial instruments and, (iii) improving and increasing domestic food production despite the constraints of resources such as water.

The agricultural sector in Oman completely depends on water availability as the Sultanate falls within the arid and semi arid belt where average rainfall is about 100 mm year^{-1} and reaches a maximum of 200 mm year^{-1} in very few regions such as Al-Jabal Al-Akhdar and Dhofar. In terms of arable land, Oman has about 2.2 million hectares (Mha) available for farming which is equivalent to 7 % of the total area of Oman, (31.4 Mha). However, the actual cropped/farmed area in the Sultanate is about 62,000 ha or 2.8 % of the total arable land (Annual Agricultural Statistics 2010). Since the year 2000, the Government spent RO 20.1 $\times$ 10^6 on agriculture and fishery development, and another RO 39.4 $\times$ 10^6 on water resources development. The investment on water resources development amounts to approximately US$1,650 per ha of cropped land. For a small country with limited potential for water resources development, this is considered a substantial investment. Based on these statistics it is clear that agricultural production in Oman is seriously impeded by land and water shortage. Under this situation, therefore, improving land and water productivity is very crucial. The existence of milder weather in regions such as Al-Jabal Al-Akhdar and Dhofar (Salalah), however presents a niche opportunity for Oman to expand production of key crops and in doing so contribute towards national efforts to achieve food security.

Food price shocks in 2008 have made countries to start rethinking about their food security situations. Most of the countries in the Gulf Cooperation Council region (GCC), Oman included, are currently putting a lot of effort towards achieving food security. Measures taken by these countries include: finding ways to increase agricultural production from within their borders; investing in other countries to produce food in order to secure food supplies to meet their needs at home and also by procuring food supplies from the world market and using these supplies to maintain a strategic food reserve. This chapter seeks to explore the potential to increase domestic production of vegetables and fruits within Oman. Specifically, this chapter looks at the potential of transforming Salalah into Oman's fruits and vegetable basket.

Salalah which is situated in the Dhofar region of the Sultanate is chosen because it has a milder weather as compared to other regions. Salalah area receives seasonal rainfall as a result of the monsoon winds from the Indian Ocean saturated with cool moisture and heavy fog. Even though the monsoon season starts late June and extends to late September, the area experiences occasional rains for a good part of the rest of the year. Salalah therefore presents itself as suitable area for expansion of vegetable production.

5.2 Past Trends in Area, Production, Imports and Exports

This section looks at trends in area, production, import and exports of fruits and vegetables into Oman. Oman has a diverse topography that range from the northern plains with variable relative humidity and temperature; mountains reaching over 3,000 m high with subzero temperatures during winter; to the summer monsoon in Salalah. As a result of this diverse topography different crops are currently grown in different parts of Oman. For example, Al-Jabal Al-Akhdar is dominated by temperate crops such as apple, pear, apricots, pomegranate and nuts. Vegetables such as tomato, lettuce, green peppers, cucumbers, cabbage are grown mostly in the Batinah region and some parts of the interior regions. Al-Rustaq area is famous for garlic production. The southern part of the country (Salalah area) is mostly grown with coconut palms, banana, papayas, and other tropical fruit trees, as well as a variety of vegetables. The rest of northern Oman is dominated by date palms and other subtropical trees such as mangoes and citrus (Lime).

5.2.1 Trends in Area and Production

Trends in area and production of fruits and vegetables in Oman from year 2000 to 2007 are as indicated in Table 5.1.

Table 5.1 shows that the harvested area has decreased over the period 2000–2007 by 8.2 %. The highest harvested area was observed in 2000 (50,853 ha) and 2004

Table 5.1 Oman horticulture harvested area and production 2000–2007

Year	2000	2001	2002	2003	2004	2005	2006	2007
Area (ha)	50,853	49,412	49,247	48,798	53,227	46,732	46,434	46,695
Yield (t ha^{-1})	10.0	11.0	9.7	9.3	9.8	10.6	10.9	10.8
Production (t)	510,762	545,368	477,009	452,965	520,978	495,423	503,865	506,500

Source: FAOSTAT (2008)

Table 5.2 Quantities and value of fruits and vegetables imported to Oman from year 2000 to 2007

Year	Imported quantities (t)	Import value (1,000 $)
2000	175,358	105,014
2001	211,999	96,100
2002	248,328	121,107
2003	282,824	140,812
2004	312,700	155,285
2005	303,405	181,434
2006	334,651	163,830
2007	332,939	186,694
Total	2,202,204	1,150,276

Source: FAOSTAT (2008)

(53,227 ha). It is not clear what triggered this downward trend in the harvested acreage. One possible explanation could be water shortage or it could also be a result of improvement in production technology. Yield per ha is more or less stagnant. Overall total production of fruits and vegetables for the period 2000–2007 remained more or less stagnant at around 500,000 t. This could be explained by the lack of technology intervention. Meaning that better use of technology e.g., the introduction of new seed varieties, efficient use of water, coupled with better crop husbandry practices should lead to increased production.

5.2.2 *Import Trends*

Based on FAOSTAT (2008) import statistics, Oman imported approximately 2.2×10^6 t of 128 different types of fruits and vegetables between 2000 and 2007. In total these imports were worth 1.15 billion US dollars as indicated in Table 5.2.

Table 5.2 shows that Oman imports of fruits and vegetables almost doubled between 2000 and 2007. Fruits and vegetable imports increased from approximately 175,000 t in 2000 to close to 333,000 t in 2007. The value of fruits and vegetables imported to Oman during the same period increased from around 105×10^6 dollars in year 2000 to 186.7×10^6 dollars in year 2007, which is about 77.8 % increase. The increase translates to an annual average of 275,276 t worth 143.78×10^6 dollars. The fifteen (15) leading fruits and vegetables imported and their tonnage are as indicated in Table 5.3.

Table 5.3 The leading fruits and vegetables imported from 2000 to 2007 in tons

		Year imported								
Rank	Commodity	2000	2001	2002	2003	2004	2005	2006	2007	Total
Vegetables										
1	Onions, dry	28,706	36,670	32,749	40,501	33,071	29,868	36,076	37,810	275,451
3	Potatoes	19,489	16,202	25,680	24,039	24,006	23,563	24,941	23,881	181,801
4	Garlic	9,548	2,224	4,928	10,933	12,021	8,681	8,414	6,985	63,734
5	Tomatoes	10,661	23,083	9,235	4,602	3,996	4,485	4,600	3,069	63,731
6	Watermelons	0	12,669	9,657	9,660	4,913	6,005	6,845	7,854	57,603
7	Cabbages	2,459	4,360	4,821	7,092	3,207	3,939	3,949	9,717	39,544
10	Lemons and Limes	496	1,057	2,663	2,208	3,733	3,447	2,406	1,966	17,976
11	Carrots and turnips	–	–	–	3,721	–	3,837	4,166	5,529	17,253
12	Cauliflowers and Broccoli	3,110	1,504	2,868	346	4,275	751	818	2,207	15,879
13	Cucumbers and Gherkins	3,072	1,439	1,172	788	1,033	1,585	1,980	512	11,581
14	Chillies and Peppers	363	355	424	262	664	746	2,779	4,693	10,286
15	Lettuce and Chicory	403	410	509	1,060	1,595	251	1,794	922	6,944
Fruits										
1	Oranges	20,745	27,955	28,303	36,933	39,990	33,193	34,568	35,515	257,202
2	Grapes	5,795	4,989	5,572	4,805	4,206	4,920	4,121	3,456	37,864
3	Bananas	1,967	1,888	2,742	2,313	4,915	2,661	4,394	5,746	26,626
Annual total (t)		**106,814**	**134,805**	**131,323**	**149,263**	**141,625**	**127,932**	**141,851**	**149,862**	**1,083,475**

Source: FAOSTAT (2008)

Table 5.4 Quantities and value of fruits and vegetables exported by Oman from year 2000 to 2007

Year	Exported quantities (t)	Export value (1,000 $)
2000	28,569	20,732
2001	24,363	21,491
2002	34,871	20,643
2003	29,897	18,707
2004	38,288	25,241
2005	35,609	26,778
2006	35,593	24,320
2007	28,527	21,314
Total	255,717	179,226

Source: FAOSTAT (2008)

Table 5.3 shows that 50 % ($1.08 \approx 1.1 \times 10^6$ t) of the 2.2×10^6 t of fruits and vegetables imported to Oman during the period 2000 and 2007 are attributed to the first 15 leading fruits and vegetables. Some of these fruits and vegetables are already commonly grown in Salalah. Table 5.3 also shows that dry onion, oranges, potatoes, garlic and tomatoes are the five leading commodities imported to Oman. Garlic and tomato imports have however decreased over the years, meaning that more of these commodities are being produced locally. Growing the 15 leading fruits and vegetables locally therefore on average has the potential to directly inject RO 143.78×10^6 to the local economy as direct income to fruits and vegetable growers. This will have significant multiplier effects in terms of job creation and increased demand of other goods and services in the local economy.

5.2.3 Export Trends

Based on FAOSTAT (2008) export statistics, Oman exported approximately 255,717 t of different types of fruits and vegetables between 2000 and 2007. In total these exports were worth 179.226×10^6 dollars as indicated in Table 5.4.

Table 5.4 shows that Oman exports of fruits and vegetables remained more or less stagnant between 2000 and 2007. Fruits and vegetable exports increased from 28,569 t in 2000 reaching the highest level of 38,288 t in 2004. Thereafter exports decreased to 28,527 t in 2007. The value of fruits and vegetables exported by Oman during the same period increased from 20.732×10^6 dollars in year 2000 to 26.778×10^6 dollars in year 2005 (the highest). Thereafter export value decreased to 21.314×10^6 dollars in 2007. The 15 leading fruits and vegetables exported by Oman between 2000 and 2007 are as presented in Table 5.5.

Table 5.5 shows that date are the leading export commodity followed by green beans, tomatoes, garlic and green chillies. Oman exports of the 15 leading fruits and vegetables decreased between 2000 and 2007. In total 137,736 t of the 15 leading export commodities were exported between 2000 and 2007.

Table 5.5 The 15 leading fruits and vegetables exported by Oman from 2000 to 2007

Commodity	2000	2001	2002	2003	2004	2005	2006	2007	Total (t)
Dates	7,711	9,881	13,019	5,025	4,691	4,752	4,080	4,097	53,256
Beans, green	2,302	1,280	3,334	2,658	2,606	3,894	2,846	3,091	22,011
Tomatoes	1,818	734	37	1,102	753	2,821	3,155	480	10,900
Garlic	5,682	438	0	85	193	267	275	91	7,031
Chillies and peppers, green	219	239	739	1,227	938	849	1,204	715	6,130
Watermelons	0	98	2,269	1,878	602	59	423	163	5,492
Cabbages	506	808	1,394	750	245	320	634	144	4,801
Oranges	118	57	266	254	875	1,580	390	937	4,477
Potatoes	328	87	369	464	1,750	251	254	947	4,450
Bananas	196	375	588	785	398	416	1,008	405	4,171
Carrots and turnips	–	–	262	653	0	1,100	1,090	864	3,969
Lemons and limes	421	448	327	123	729	418	501	247	3,214
Onions, dry	1,731	1,057	57	23	21	66	60	66	3,081
Mushrooms and truffles	144	330	69	28	162	278	857	594	2,462
Cucumbers and gherkins	100	293	1,074	460	152	89	120	3	2,291

Source: FAOSTAT (2008)

5.3 Is It Feasible to Expand Fruits and Vegetables Production in Salalah (Dhofar)

This section examines whether it is feasible to expand fruits and vegetable production in Salalah. Fruits and vegetables being proposed are as indicated in Table 5.3. To establish whether it is feasible to expand the production of these fruits and vegetables we need to look at, (1) land and water availability in Salalah, (2) weather and soil requirement for the proposed crops, and (3) the gross margins for these crops in order to ascertain whether it makes economic sense to expand the production of these commodities.

5.3.1 Land and Water Availability

In this section we look at land and water requirements and availability. We start by presenting data on yield per ha for the 15 crops proposed (Table 5.6) and based the yield data we compute the amount of land required based on the imported quantities from Table 5.3. From Table 5.3 we used the highest recorded import quantity to compute the amount of land needed (see Table 5.7).

Table 5.7 shows that to grow the 15 leading imported fruits and vegetables in Salalah we need approximately 16,610 ha. Salalah has over 21,000 ha of cultivated lands at present (FAO 1992).

Table 5.6 Fruits and vegetables comparative yields (t ha^{-1})

Crops	Countries			
	Oman	Saudi Arabia	India	Pakistan
Onions, dry	20.45	23.69	10.16	13.16
Oranges	NA	NA	8.74	12.72
Potatoes	28.42	23.54	19.29	19.03
Garlic	9.77	NA	4.39	7.85
Tomatoes	45.22	32.31	17.95	10.09
Watermelons	30.94	20.21	12.75	14.75
Cabbages and other brassicas	26.58	NA	22.69	16.51
Grapes	NA	12.37	26.04	7.24
Bananas	11.87	NA	35.87	4.43
Lemons and limes	4.86	NA	8.48	12.75
Carrots and turnips	10.29	17.21	14.58	17.66
Cauliflowers and broccoli	22.14	NA	17.99	18.03
Cucumbers and gherkins	25.04	64.78	6.67	5.97
Chillies and peppers, dry	17.66	NA	1.63	1.97
Lettuce and chicory	NA	NA	6.58	1.00

Source: FAOSTAT (2008)

Table 5.7 The amount of land required to grow the 15 leading imported commodities

	Crop	Yield in tons per ha	Highest recorded imports in tons	Land required in ha
Vegetables				
1	Onions, dry	20.45	40,501	1,980.5
3	Potatoes	28.42	25,680	903.6
4	Garlic	9.77	12,021	1,230.4
5	Tomatoes	45.22	23,083	510.5
6	Watermelons	30.94	12,669	409.5
7	Cabbages	26.58	9,717	365.6
10	Lemons and limes	4.86	3,733	768.1
11	Carrots and turnips	10.29	5,529	537.3
12	Cauliflowers and broccoli	22.14	4,275	193.1
13	Cucumbers and gherkins	25.04	3,072	122.7
14	Chillies and peppers, green	17.66	4,693	265.7
15	Lettuce and chicory	1.00	1,794	1,794.0
Fruits				
1	Oranges	8.74	39,990	4,575.5
2	Grapes	12.37	5,795	468.5
3	Bananas	11.87	5,746	484.1
Total amount of land in ha required				16,609.0

5.3.2 Weather, Water and Soil Requirements for the Proposed Crops

Environment of a certain site has the potential to affect crop production through changes in temperature, soil, water, rainfall (timing and quantity), solar radiation and the interaction of all these elements. Environmental factors have profound effect on plant phenotype and determine the extent to which genotype potential expressed. An understanding of the response of crops to environmental variables is crucial to minimize the deleterious impact of suboptimal environmental conditions and to manage crops for maximum productivity. Table 5.8 explains important variables of environment e.g. temperature, water and soil requirements for the proposed 15 fruits and vegetable crops.

Table 5.8 shows that with the exception of oranges, grapes, lemons and limes, the rest of the crops require temperatures ranging from 18 to 25 °C. This temperature range is within the seasonal temperature averages for Salalah as indicated in Table 5.9.

A close look at Tables 5.8 and 5.9 indicate that on average the temperature range in Salalah is from 22 to 25 °C. This means that, onions, garlic, tomatoes, watermelons, bananas, cucumbers and chillies can be comfortably grown. Other crops such as potatoes, cabbages, carrots, cauliflowers and broccolis and lettuce can also be comfortably grown even though their temperature range is 4 °C lower than that of Salalah. We can therefore conclude that Salalah temperatures are conducive for growing the proposed crops.

Some of the crops require sandy loam and others light well drained soil with a pH range of 5.5–7 (see Table 5.8). Salalah plains consist of two major soil orders, the Aridisols and Entisols. The Aridisols, dominated by the Calciorthids, are loamy-skeletal, deep alluvial soils. These alluvial soils occur on nearly level to gently slopes, calcareous, and with moderate hydraulic conductivity. These soils are suitable for large-scale irrigated farming and the major limitation is the high gravel content and of course recently soil salinity is another problem. The Entisols, dominated by Torrifluvents, which are also loamy, deep soils. These are alluvial soils with appreciable amounts of $CaCO_3$ accumulations and more of silt. Compared to the Calciorthids these soils are having very low gravel content, and much more suitable for agriculture. The pH of the soil is about 7.8, hence not a constraint to the crops identified. In summary, Salalah soils are well drained, and the pH is marginally high for crops identified. However, limitations of pH can be managed through better nutrient and fertilizer management.

5.4 Conclusions

Primary crop production in Oman in 2005/2007 was 486.872 metric tons of which contribution of fruits and vegetables were 353,072 metric tons and 102,606 respectively. In comparison, only 26,206 metric tons of cereals were produced. The value

Table 5.8 Environmental demand for the 15 leading fruits and vegetables

Nos	Vegetable types	Temperature °C (optimum range)	Soil requirements	Water requirements (m^3 ha^{-1} $year^{-1}$)
1	Onions	20–25 (insensitive to frost)	Fertile and well drained, pH 6–6.8	283.6 Require even moisture throughout growing season
2	Oranges	13–35 (sensitive to frost)	Sandy loam good drainage, pH 6.5–7.5	28,230 Drought & salinity sensitive
3	Potatoes	18–25 (sensitive to frost)	Good drainage, pH 4.8–7.5	Drought sensitive
4	Garlic	20–25 (Insensitive to frost)	Fertile and well drained, pH 6–6.8	Soils must not be dried excessively
5	Tomatoes	25–27 (sensitive to frost)	Good drainage, pH 5.8–7, can tolerate salinity upto 7 dSm^{-1}	269.5 5–7 days interval
6	Watermelons	25–27 (sensitive to frost)	Light with good drainage, pH 6–6.8	462.1 Sensitive to water stress
7	Cabbages	18–25 (insensitive to frost)	Can be grown on variety of soils (sandy loam best), pH 5.5–6.8	271.1 At least 2.5–3.8 cm of water per week, does not like erratic moisture
8	Grapes	15–35 (sensitive to frost)	Well drained, pH 6.5–7.5	5–7 days interval
9	Bananas	20–30 (sensitive to frost)	Good drainage, sensitive to salinity above 0.8 dS m^{-1}	37,810
10	Lemons and limes	13–35 (sensitive to frost)	Sandy loam good drainage, pH 6.5–7.5	Drought & salinity sensitive
11	Carrots and turnips	18–25 (insensitive to frost)	Deep uniform light textured soil, pH 5.5–7	277.6 Uniform soil moisture, 3.8 cm water weekly or more in arid
12	Cauliflowers and broccoli	18–25 (insensitive to frost)	Well-drained sandy loam, pH should not fall below 5.5	Needs enough moisture in dry season
13	Cucumbers and gherkins	20–25 (sensitive to frost)	Light soil, good drainage, pH 6.0–6.8, below pH 5.5 can reduce fruit set	247.3 3.8–4 cm of water is required, very sensitive to drought
14	Chillies and peppers	25–27 (sensitive to frost)	Sandy loam, pH 5.5–7	Sensitive to water stress
15	Lettuce and chicory	18–25 (insensitive to frost)	Sandy loam and clay loam soil, pH 6.5–7	Need large amount of water, sensitive to drought

Sources: Krug (1991) and Morton (1987)

Table 5.9 Salalah seasonal weather averages (monthly normals)

	Monthly normals												
	Jan	Feb	Mar	Apr	May	June	July	Aug	Sept	Oct	Nov	Dec	Average
High temp (°C)	26	27	28	31	31	31	27	25	28	29	29	28	28
Low temp (°C)	18	19	21	24	26	27	24	23	23	22	21	20	22
Precipitation (mm)	2.2	7.0	6.3	19.8	17.1	10.6	24.6	24.5	4.1	4.1	9.6	1.1	11
Avg. % sunshine	40	36	41	43	47	28	6	6	26	44	42	41	33

Source: http://www.wunderground.com/NORMS/DisplayIntlNORMS.asp?CityCode=41316&Units=metric

of production of cereals and vegetables were 7.8×10^6 and 17.6×10^6 RO respectively. This comparison confirms that Omanis prefer producing high value vegetables to cereal crops. Oman imported approximately 2.2×10^6 t of 128 different types of fruits and vegetables between 2000 and 2007. This chapter evaluated land, water, climate and soils constraints to producing 15 leading imported fruits and vegetables in Salalah. The plains consist of 21,000 ha of cultivable lands, and assuming two season per year, conversion of 8,000 ha into vegetable lands will be sufficient to produce the 15 leading fruits and vegetable imported. The alluvial aquifer receives 54 MCM water as recharge and almost all of it is used for irrigation. Hence, the groundwater levels are somewhat steady in recent years. The minor crop irrigated is fodder, which has an actual evapotranspiration equal to that of the potential evapotranspiration. Hence changing the land use from fodder to vegetables will not increase demand for irrigation water demand. Soils in the plains are deep and well-drained and the pH is closer to neutral. The weather, especially temperature ranges between 15 and 32 °C. The coolest nights are in winter and the warmest at day time in summer. Hence, all leading imported crops can be grown in this climate.

In summary Salalah's climate, soils, and land and water availability are conducive to grow adequate fruits and vegetables. This will improve on-farm income, and possibly lead to value chain opportunities. However its current land use is dominated by fodder for animals, and in general water management practices is highly inefficient. Hence it may be prudent to investigate the potential to transform Salalah into Oman's vegetable basket, considering other factors such as technological advancements, and socio- economics.

References

Annual Agricultural Statistics (2010) http://www.maf.gov.om/Pages/PageCreator.aspx?lang=EN&I=0&CId=0&CMSId=800631&DId=0

FAO (1992) The Oman soil atlas. www.fao.org/ag/AGP/AGPC/doc/Counprof/oman/oman.htm

FAOSTAT (2008) http://www.fao.org/corp/statistics/en/. Accessed in May 2010
http://www.nhb.gov.in/Horticulture%20Crops/Grape/Grape1.htm. Accessed in May 2010
http://www.wunderground.com/NORMS/DisplayIntlNORMS.asp?CityCode=41316&Units=metric. Accessed in May 2010
Krug H (1991) Gemuseproduktion, 2nd edn. Verlag Paul Parey, Berlin, 541 pp
Morton JF (1987) Fruits of warm climates. Creative Resources System, Winterville (Banana)
World Bank (2009) Improving food security in Arab countries. The World Bank, Washington, DC

Chapter 6
Climate Change, Water Scarcity and Food Security Complex: A Case Study from Bahrain

Salma Saeed Ahmed Bani

Abstract Food production is a complex and interrelated system. It operates within complex systems and is multifunctional in its nature. The complex connections between water, food security and climate change is likely to create negative impact on food production and this will remain a crucial policy issue. Due to hyperarid climatic conditions, the Kingdom of Bahrain is characterized by high temperatures, erratic and often scanty rainfall, high evapotranspiration rates and high humidity levels due to the surrounding Gulf waters, as a result of which water in Bahrain is inherently scarce. Water scarcity will reduce agricultural production and threaten country food security. Bahrain requires a reallocation of domestic resources in order to increase agricultural production and boost the contribution of agricultural sector to its Gross Domestic Product (GDP). To reduce the deficit between local food production and imports, Bahrain needs to achieve relative food security relying on local production of certain strategic items and to encourage agricultural investment and optimize the role of the private sector in developing the sector. Bahrain should also increase investment in research and development in agriculture sector. Thus, answers to the resolving issues of food security and stability in food production will increasingly come from improved water management, increased water use efficiency and its sustainable use in the agriculture sector.

Keywords Bahrain • Climate change • Food production • Food security • Water scarcity

S.S.A. Bani (✉)
Planning and Municipal Council Department, Ministry of Municipality and Urban Planning, Manama, Bahrain
e-mail: Salmabani@yahoo.com

S.A. Shahid and M. Ahmed (eds.), *Environmental Cost and Face of Agriculture in the Gulf Cooperation Council Countries: Fostering Agriculture in the Context of Climate Change*, DOI 10.1007/978-3-319-05768-2_6, © Gulf Research Centre Cambridge 2014

6.1 Introduction

The Kingdom of Bahrain is an island nation in the Arabian Gulf and consists of an archipelago of 36 low-lying islands. Bahrain is one of the Gulf Cooperation Council (GCC) member countries including Kuwait, Oman, Qatar, Saudi Arabia, and United Arab Emirates.

Arabian Peninsula is located in the southwestern region of the Asian continent. Covering about 3×10^6 km^2, the southeastern area of the peninsula is the Rub'al-Khali, the Empty Quarter, which is the world's largest expanse of continuous sand. Politically, the Arabian Peninsula consists of Saudi Arabia, Kuwait, Bahrain, Qatar, the United Arab Emirates, the Sultanate of Oman, and the Republic of Yemen. Together, these countries (excluding the Republic of Yemen) constitute the Gulf Cooperation Council (GCC) countries. Founded on 26 May 1981, the aim of this collective is to promote coordination between member states in all fields in order to achieve unity.

Bahrain (from the Arabic word for "two seas"), is believed to have separated from the Arabian Peninsula around 6000 BC. Located in the Arabian Gulf (Fig. 6.1), the islands are about 24 km from the east coast of Saudi Arabia and 28 km from Qatar. The total area of the islands is about 678 km^2.

Bahrain got independence from UK on 15 August 1971. Bahrain is the smallest of the GCC countries. It is located 26 00 N and 50 33 E. It has 100 % boundaries to the seas (0 % land boundaries with neighboring countries). Its coastline is stretched

Fig. 6.1 Location of the Kingdom of Bahrain (Source: http://www.lonelyplanet.com/maps/middle-east/bahrain/. Last accessed 26 Feb 2013)

over 161 km. Its climate is arid; mild, pleasant winters; very hot, humid summers. Its terrain is mostly low desert plain rising gently to low central escarpment. Bahrain has only 2.82 % arable land. Irrigated land is about 50 km^2. Periodic droughts and dust storms are the most prevalent natural hazards. There is general lack of surface water (fresh water resources); ground water and seawater are the major sources to meet the demands of public, agriculture and industrial sector. Bahrain has 157 m^3 $year^{-1}$ total annual renewable water resources (TARWR) and total water use is 258 % of TARWR (AQUASTAT and FAO 2005).

To join the international community and to conserve natural resources, Bahrain has signed various agreements (Biodiversity, Climate change, Desertification, Hazardous Wastes, Law of the Sea, Ozone Layer Protection, and Wetlands). Total population of Bahrain is 688,345 including 235,108 expatriates (July 2005 estimates) and 1.5 % population growth rate per annum. In well-to-do Bahrain, petroleum production and refining account for about 60 % of export receipts, 60 % of government revenues, and 30 % of GDP.

Major agricultural products are fruits, vegetables, poultry, dairy products; shrimp, fish. Agriculture contributes to 0.7 % national GDP. Major industries are petroleum processing and refining, aluminum smelting, iron pelletization, fertilizers, offshore banking, ship repairing; tourism. In 2004, its import partners were (Saudi Arabia 32.4 %, Japan 7.3 %, Germany 6.1 %, US 5.6 %, UK 5.4 %, France 4.8 %).

6.2 Climate of Bahrain

Due to arid climatic conditions the country is characterized by high temperatures, erratic and often scanty rainfall, high evapotranspiration (with peaks of more than 10 mm day^{-1}) rates and high humidity levels due to the surrounding Gulf waters. Temperature averages from 17 °C in winter (December–March) to 35 °C in summer (June–September). The rainy season runs from November to April, with an annual average of 83 mm, sufficient only to support the most drought resistant desert vegetation.

6.3 Soil Resources of Bahrain

The soils of Bahrain are mostly moderate to shallow in depth. The topsoil texture ranges from sand to loamy sand whereas subsoil texture varies from loamy sand to sandy loam. The water holding capacity is very low and the available moisture is about 2–6 %. Infiltration rates are very high, above 120 mm h^{-1}. Most of the cultivated land became saline, mainly due to heavy applications of saline water during irrigation.

6.4 Agricultural Land

Agriculture development is concentrated on the north and northwest coast (Fig. 6.2), where this is due to potentially suitable soils, water quality and availability. In the past, springs located at the contact of the limestone uplands (Dammam black slope) and the coastal fringe deposits were used for irrigation on the coastal lowland soils. These fairly large continuous areas of flat and permeable soils are served by groundwater of moderate quality, and this zone has been intensively cultivated. The land use of Bahrain is shown in Fig. 6.3, which clearly illustrates major part of the land is a wasteland and only a fraction is used for irrigated farming.

The total arable land in Bahrain is estimated to be 64,000 donums in other words it is about 10 % of the total area which amounts to 622 km^2. Two thirds of this arable land is cultivated.

Agriculture in the Kingdom of Bahrain witnessed in recent years many obstacles that affected its role in the development process and achieving food security in the country. The value of agricultural output is 16.2×10^6 dinars at a contribution rate of 23 % of the GDP, and the value of food imports amounted to more than 202×10^6 Bahraini dinars, and the deficit of the balance of commodity trade in the Kingdom of Bahrain is up to 173×10^6 dinars. Therefore, with the world facing perfect storm of food scarcity, Bahrain needs to focus on lowering its food imports and increasing local agricultural production in order to boost the contribution of agricultural sector to its GDP. Bahrain, in order to overcome the deficit between food production and imports need to achieve relative food security relying on local production of certain strategic items and to encourage agricultural investment and

Fig. 6.2 Agriculture farm in Bahrain

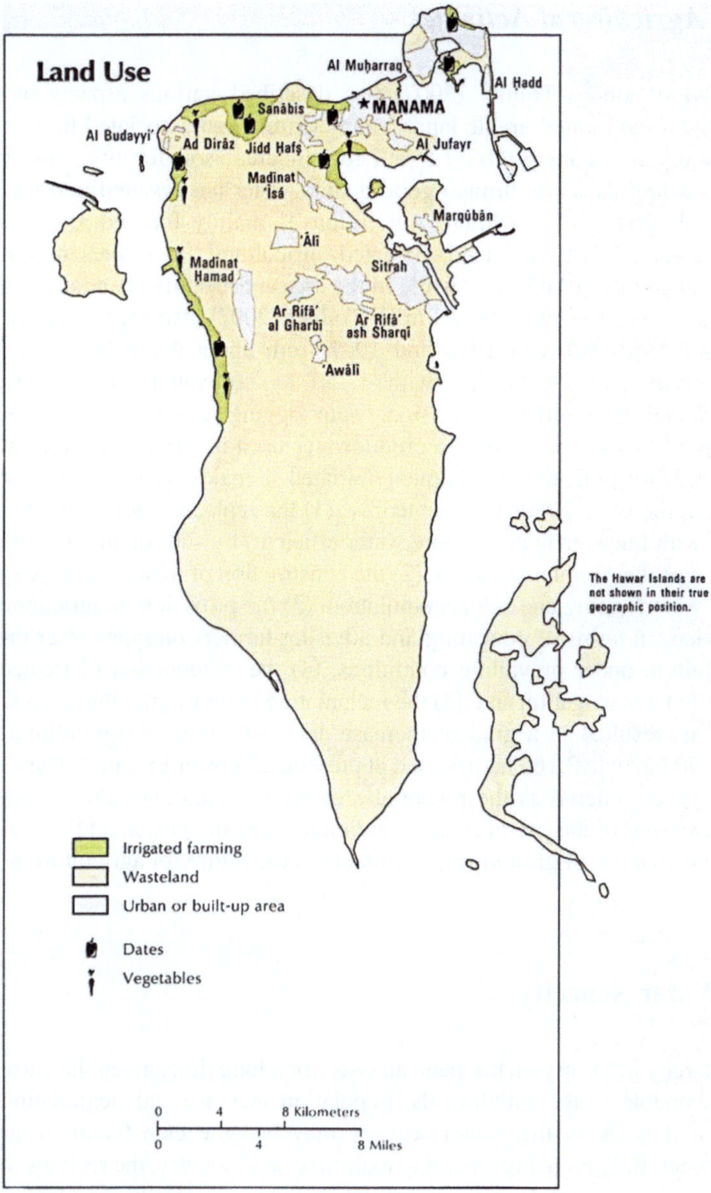

Fig. 6.3 Land uses in Bahrain

optimize the role of the private sector in developing the sector. Thus productivity and sustainability of the Bahrain food system, especially under more severe climate change scenarios is a major concern for the government. The government applies the economic incentive strategy to enhance crop productivity.

6.4.1 Agricultural Activities

In a previous study, Tollner (2007) has described various aspects of Bahrain. Accordingly, the limited arable lands, sandy texture and associated high infiltration rate, low organic matter (0.05–1.5 %), low inherent soil fertility, low water and nutrient holding capacity, limited good quality water has resulted into focused low agricultural activities in Bahrain. Agriculture is mainly focused along the north-western coast of Bahrain Island. Irrigated agricultural farms present soil salinity within a range of 4–12 dS m^{-1}, while in the areas of recently abandoned agriculture (1,065 ha) it could reach to 60 dS m^{-1}. Tollner (2007) also expressed declining of agricultural lands between 1956 and 1977 from about 6,460 ha (with 3,230 ha cultivated) to about 4,100 ha (with 1,750 ha cultivated). This decrease was attributed mainly to urban expansion, waterlogging and soil salinization due to deterioration of the quality of the groundwater used in irrigation. In an attempt to reverse the situation, the government initiated a major agricultural development program in the early 1980s represented by, (1) the replacement of surface irrigation methods with micro-irrigation (more water efficient) by subsidizing more than 50 % of the cost of their implementation, (2) the construction of major drainage systems to alleviate waterlogging and salt accumulation, (3) the provision of agricultural extension services in terms of educating and advising farmers on types of crops suitable for agriculture under prevailing conditions, (4) the introduction of treated sewage effluent (TSE) in irrigation and, (5) the reclamation of new agricultural lands (Tollner 2007). This resulted in a gradual increase and restoration of agricultural lands to about 4,230 ha, with 3,165 ha irrigated at present, all power irrigated. These 4,230 ha can also be considered as the potential area for irrigated agriculture, should there be an increasing future use of nonconventional water sources, in addition to groundwater. The quantity of groundwater available in the future for agriculture is difficult.

6.5 Water Scarcity

Water scarcity in the region has been an issue for a long time, given the current trends of unsustainable water withdrawals, population increase and degradation of land resources; it is likely that water scarcity may become cute factor in sustainable development. Bahrain being an arid to extremely arid country, the recharge of aquifer is very slow or not at all. Groundwater has become less accessible and less acceptable environmentally; therefore, sufficient availability and adequate water quality are of crucial importance for sustainable development and protection of the environment. The question, however, is pertinent that can we increase water productivity and ensure enough water for sustaining the resource base for food production?

Water scarcity will reduce agricultural production and threaten country food security. Therefore, strategies to optimize water use in agriculture under conditions of scarcity need be developed to maximize return per unit of water instead of per unit of land and to improve local livelihoods.

6.6 Availability of Water Resources and Uses

Agriculture under greenhouse (protected agriculture) was introduced in 1976 with the aim of increasing production and achieving a higher level of self-sufficiency in various agricultural products, particularly high-quality fresh vegetables crops. The main greenhouse crops produced are: tomato, cucumber, pepper, squash, eggplant, lettuce, strawberry, bean and cauliflowers. However, new policies and institutions are needed for implementing a sound water use development program under these conditions.

6.6.1 Water Used for Agriculture

The demand for water in Bahrain comes from domestic, agricultural and industrial sectors. The water demands in Bahrain are met through groundwater desalinated water and treated sewage effluent. About 70 % of the total water demand is met by the island's groundwater resources. Unfortunately due to the lack of rain, agriculture is irrigated and mainly depends on ground water; therefore the agriculture sector is the main groundwater consumer and consumes about 80 % of the groundwater abstracted. The need to reduce groundwater abstractions has prompted the Government to consider the use of treated sewage effluent as an additional source of water for agricultural purposes. However, it is not utilized to its full capacity, and only 20 % of the treated effluent is used, mainly on experimental farms, landscaping and certain industrial uses.

6.6.2 Use of Alternate Water Sources

The productivity and sustainability of the Bahrain food system, especially under more severe climate change scenarios is a major concern for the government. The government applies the economic incentive strategy to enhance crop productivity. Effort is being made in the agricultural sector on adoption of sustainable development agricultural policies; to promote improved water use efficiency in agriculture and the utilization of treated waste water (TWW). The positive point of increasing the use of TWW is in the fact that its quantity will increase with increasing population. The beneficial component of TWW use is of its content on nutrients (N-P) required for crop growth. In addition, the use of biosolids and soil amendments will improve the soil quality and crop production. The challenge for the twenty-first century in agriculture is water. Indeed, it is expected that in the coming 30 years, increase in production will come mainly from enhancing water productivity. Meaning, water use efficiency and use of marginal waters including waste waters. The conditions are more acute in MENA and GCC countries including UAE.

Globally, the two main drivers of waste water use are water pollution and water scarcity. Around the world many countries are using TWW for irrigation. For instance, the agricultural area under application of TWW reaches over 40,000 ha in Egypt, around 22,000 ha in Argentina, 17,000 ha in UAE and 15,000 ha in USA.

The use of wastewater in agriculture has long been of interest of scientists and professionals. Recently number of studies have been published on the efficient management of wastewater, its treatments and reuse in water scarce countries (Al Baz et al. 2008), including a Review (Integrated wastewater management) by Abbasi and Al Baz (2008), effect of pathogens in tomato plants (Halalsheh et al. 2008), socio-economic aspects of wastewater treatment and water reuse (Sheikh 2008) and wastewater reuse for agriculture pilot project in Jordan (Al-Ghazawi et al. 2008).

While the use of TWW is beneficial for soil and crop, the issue of pathogens and contamination has to be explored. In fact, Shuval et al. (1986) pointed out that pathogen contamination are only detected in association with the use of raw or poorly-settled wastewater, while inconclusive evidence suggested that the appropriate wastewater treatment could provide a high level of health protection. In this context it is believed that in Bahrain wastewater is treated to tertiary level and hence reduces health risks to a significant extent (Shuval et al. 1985, 1986).

New policies and institutions are needed for implementing a sound agricultural water use development program under these conditions.

Also, to ensure food security since 1970s of last century, government of Bahrain become responsible for providing basic food commodities to the nation, but since 2001 government opened doors for local merchants to import, store, distribute and sell these commodities in the local market. Besides, for food security purpose Bahrain government has established safeguard mechanisms for three strategic food commodities, which are: flour, imported Australian meat and locally produced chicken.

6.6.3 Desalinated and Treated Wastewater

In 1991, the total quantity of desalinated water used was 44.1×10^6 m^3 year^{-1}. In addition Bahrain treats about 45×10^6 m^3 year^{-1} of wastewater (secondary treatment). Only 8×10^6 m^3 year^{-1} receives tertiary treatment and are used for irrigation purposes in government farms and some private farms, while the rest is discharged to the sea. The chemical and hygienic properties of the tertiary treated water are within international limits and are considered good for agricultural purposes. Although government plans for full utilization of the TSE (Treated Sewage Effluent) through major agricultural projects, delay and lack of finances for these projects have caused limitations in the use of these waters. About 94 % of the water used in agriculture, including livestock, is groundwater and 6 % is TWW, while for domestic and industrial purposes about 60 % of the water used is groundwater and the remaining part desalinated water. Available and exploitable water resources are shown in Table 6.1.

Table 6.1 Available and exploitable water resources 2010 (in million cubic meters)

Water resources	Percentage of use	Exploitable resources	Available resources
I- Renewable water resources:			
(1) Surface water		–	–
(2) Groundwater		113.90	126.60
Percentage of exploitable water resources from total renewable resources	90		
Percentage of exploitable water resources for agriculture from total renewable resources.	58	73.90	
II- Non-renewable water resources:			
(1) Desalinated water	100	167.80	167.80
(2) Treated sewage effluent	100	40.15	40.15
(3) Agricultural drainage water	0.82	0.18	22.00
Total available water resources			256.55
Total Exploitable Water resources		322.05	

Source: Agricultural Engineering and Water Resources Directorate. Ministry of Municipal Affairs and Urban Planning Agriculture Affairs, Kingdom of Bahrain

6.6.4 Trends in Water Resources Management

In the past many years, the government has been taking several steps and courses of action to provide solutions to the water crisis in the country and agricultural sector deterioration. These include: water conservation campaigns in all sectors, water pricing in the domestic sector and more reliance on non-conventional water sources (TSE in agriculture and desalinated water for domestic purposes). Government policy with regard to water use is to reduce groundwater dependency for the domestic water supply, the second main water user, by constructing additional desalination plants with a total capacity of $50 \times 10^6\ m^3\ year^{-1}$. Groundwater is planned to be exclusively used for irrigation. Additional requirements for future agricultural development would be supplemented by TSE and the government is planning to increase the TSE utilization volume to about $49\ 10^6\ m^3\ year^{-1}$. The additional volume would be used to irrigate an area of about 1,810 ha using modern techniques. This will bring the total area to be irrigated by TSE to 2,240 ha, leaving the remaining area to be irrigated from groundwater (Tollner 2007).

6.7 Food Production in Bahrain

Bahrain was one of the richest countries in the Arabian Gulf even prior to the discovery of oil resources in 1932. Its' pearl was the famous and best in the region, an important agriculture and trading center. Urbanization expansion as well as industrial sector land consumption resulted in pressure on agriculture in Bahrain. The biggest challenge is limited agricultural lands and shortage of water resources.

The total arable land in Bahrain is estimated to be 64,000 donums, in other words it is about 10 % of the total area which amounts to 622 km^2. Two thirds of this arable land is cultivated.

The agriculture products produced locally covers only 12 % of total consumption needs. The major crops grown are dates and fruit trees with a yield of 7.5 t ha^{-1}, vegetables, mainly tomatoes, with a yield of 11.7 t ha^{-1}, and fodder crops, mainly alfalfa, with a relatively high yield of 74.5 t ha^{-1}. The Alfalfa tolerates high salinity and is a cash crop grown all year round with high local demand. However, because of the very high irrigation water requirements of alfalfa, it is expected that this trend will have negative implications for the country's groundwater resources. The Government assists agricultural producers mainly by offering subsidies for a number of inputs, such as pesticides, veterinary drugs, machinery services, and irrigation material.

6.8 Climate Change, Water Scarcity and Food Security Complex

Sustainable Food production is a complex and interrelated system. It operates within complex systems and is multifunctional in its nature. For example we cannot simply maximise production, without also ensuring that the system which delivers those increased yields meets society's other needs. Therefore, the complex connections between water, food security and climate change will create significant negative impact on food production and remains a crucial policy issue (Fig. 6.4).

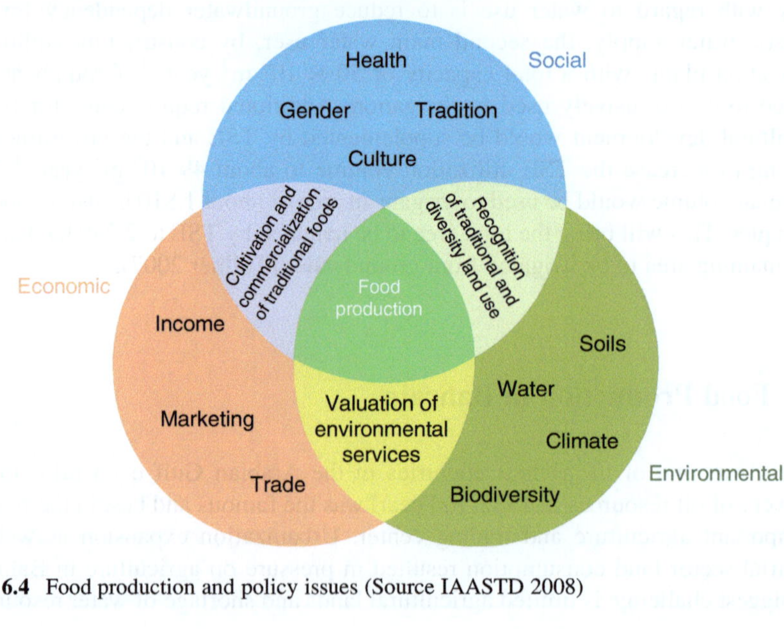

Fig. 6.4 Food production and policy issues (Source IAASTD 2008)

Water scarcity will reduce agricultural production and threaten country food security. Therefore, strategies to optimize water use in agriculture under conditions of scarcity need to be developed to maximize return per unit of water.

6.9 Climate Change and Adaptation

Climate change has been identified as the biggest environmental challenge facing the world. Technologies and practices do exist, or have been developed in different parts of the world to mitigate the impact of climate change through adaptation. Improved analysis of adaptation technologies is required to show how they can contribute to building adaptive capacity and resilience in the agriculture sector. This information needs to be compiled and disseminated for a range of stakeholders from local to national levels. The adaptations range from improved weather forecasts to water conservation, drip irrigation, sustainable soil management, better livestock management, and change in crop types and planting, among others (Clements et al. 2011). Some of these measures may need investment while the others primarily require improving awareness and building capacity to deal with new practices.

Changing weather patterns will not only have negative impact on habitat and species distribution but will drastically alter the way food can be grown in the world. There is a complex connection between water, food security and climate change. The Intergovernmental Panel on Climate Change (IPCC) anticipates rising temperature, ice melting and subsequent sea level rise, declining rainfall, soil moisture, an increase of evaporation in the region. Without changes in policy and technology, water scarcity will reduce agricultural production and threaten regional food security.

Climate change affects agriculture and food production in a complex ways. It affects food production directly through changes in agro-ecological condition and indirectly affects growth and distribution of incomes, and consequently high demands of agricultural produce. The complex connections between water, food security and climate change, will ultimately create significant negative impact on food production. Therefore, there is an urgent need to engage sustainable agricultural practices and policies to form part of the solution to current environmental change and challenges, which cannot be over emphasized.

Bahrain's environment is hostile to agriculture, characterised by extreme heat, water scarcity and high soil salinity, as a result of which domestic production is insufficient to offset the current local food requirements. Therefore, food security is an important issue for the country, and government of Bahrain has taken initiatives to enhance domestic production and at the same time to secure food imports through international agricultural investments.

6.10 GCC and Food Security Investments in Foreign Countries

Food security is a major concern of the GCC countries at the present time, due to the agricultural sector's consumption of the greatest proportion of water in these countries, and pumping money and providing support to the production of local food. This issue called on GCC governments to invest in farmland abroad through purchasing or renting, aiming at ensuring food resources in the region, especially that the GCC countries import 90 % of its food demands. In this regard, the Kingdom of Saudi Arabia is leading the trend toward investing US$23.1 billion in food security initiatives, followed by UAE and Qatar, which allocated $18.3 and 5.1 billion, respectively, to ensure food security.

6.11 Sustainable Food Security

For an overall national economic development, the strong link between food, agriculture, and people is highly important. The concept of sustainable food security combines above three elements in to a major objective that is fundamental to economic development. The issue of food security has become a major concern, on current global trends, particularly skyrocketing prices of basic commodities. Achieving sustainable food security will require more than improving farm productivity and profitability while minimizing environmental impacts. The concept is broader than sustainable agriculture; it aggregates the goals of household food security and sustainable agriculture. Despite aiming for food security since a long time, Bahrain is able to produce only a quarter of the total food demand due to unfavourable climatic conditions and limited availability of arable land. As a result of which the high dependence on imports is going to continue and this makes the issue of food security critical for the country. The governments need to undertake the necessary steps to secure imports for the growing population. Large food imports have significant consequences on financial flow to other countries, and a sense of food insecurity during political, energy, and food crises in food exporting countries. Generally, the whole world is facing food scarcity problems to various extents. Bahrain should invest in research and innovation in agriculture sector in an attempt to lowering food imports and increasing local agricultural production, to increase share of local food production to total food demand, and hence boosting the contribution of agricultural sector to its Gross Domestic Product (GDP).

6.11.1 Strategies for Sustainable Food Security

Peoples are food secure when they have regular access either through production or purchasing power to sufficient food for healthy and productive life. Therefore, from the food security perspective, it is important that sufficient food is available in the country; it is easily accessible, and affordable to all residents.

Bahrain government along with other GCC countries is currently pursuing a strategy to secure their food supply. One of the option is the investment in agriculture in countries which are eco-creditors and not eco-debtor, and have sufficient arable lands and water resources, as well as favorable climate. Other option is to encourage domestic food processing industry so that imports of processed food are decreased. This is possible through securing domestic and foreign raw materials and processing in the region. To build strategic food reserves along the lines of energy reserves in the US and food stockpiles in India.

To ensure food security, Bahrain is undertaking various initiatives, such as it is encouraging the private sector to produce leafy vegetables in greenhouses (protected agriculture). It is also supporting the private sector to produce fish, poultry, sugar and dates. Both the government and private sector firms are investing in the overseas markets to acquire farmland for the production of fruit, vegetables, rice and corn in an attempt to meet the rising demand. Bahrain has invested in farmland in India, Pakistan, Philippines, Thailand, and Turkey and Sudan.

Bahrain, like the other GCC countries, imports over 90 % of its food requirements for consumption, as a result of which food imports in the kingdom of Bahrain stood at US$1.7 billion in 2010. High dependence on imports makes the country food supply very vulnerable and highly dependent on the world food market. Disruption in food imports, either due to policy restrictions by exporting countries or natural calamities has affected the region significantly. Therefore, Bahrain, in order to overcome the deficit between food production and imports need to achieve relative food security relying on local production of certain strategic commodities and to encourage agricultural investment and optimize the role of the private sector in developing the sector.

6.12 Conclusion

The food security coupled with the water scarcity of the many developing nations is a cause of serious concern. Bahrain, in order to overcome the deficit between food production and imports need to achieve relative food security relying on local production of certain strategic items and to encourage agricultural investment and optimize the role of the private sector in developing the sector. Therefore, Bahrain requires a reallocation of domestic resources in order to increase agricultural production lowering its food imports and boost the contribution of agricultural sector to its Gross Domestic Product. Thus, answers to the resolving issues of food security and stability in food production will increasingly come from improved water management and its sustainable use in the agriculture sector.

References

Abbasi B, Al Baz I (2008) Integrated wastewater management: a review. In: Al Baz I, Otterpohl R, Wendland C (eds) Efficient management of wastewater. Its treatments and reuse in water scarce countries. Springer, Berlin/Heidelberg, pp 29–40

Al Baz I, Otterpohl R, Wendland C (eds) (2008) Efficient management of wastewater. Its treatments and reuse in water scarce countries. Springer, Berlin, 303pp

Al-Ghazawi Z, Amayreh J, Rousan L, Hijazi A (2008) Wastewater reuse for agriculture pilot project at the Jordan University of Science and Technology. In: Al Baz I, Otterpohl R, Wendland C (eds) Efficient management of wastewater. Its treatments and reuse in water scarce countries. Springer, Berlin/Heidelberg, pp 283–297

AQUASAT, FAO (2005) Water availability information by country. www.fao.org/nr/water/aquastat/countries_regions/bahrain/index.stm

Clements R, Haggar J, Quezada A, Torres J (2011) Technologies for climate change adaptation – agriculture sector. UNEPO 2011 report. Energy, climate and sustainable development, 200 pp

Halalsheh M, Ghunmi LA, Al-Alami N, Fayyad M (2008) Fate of pathogens in tomato plants and soil irrigated with secondary treated wastewater. In: Al Baz I, Otterpohl R, Wendland C (eds) Efficient management of wastewater. Its treatments and reuse in water scarce countries. Springer, Berlin/Heidelberg, pp 81–89

IAASTD (2008) Global report. International assessment of agricultural knowledge, Science and Technology for Development, Johannesburg

Sheikh B (2008) Socioeconomic aspects of wastewater treatment and water reuse. In: Al Baz I, Otterpohl R, Wendland C (eds) Efficient management of wastewater. Its treatments and reuse in water scarce countries. Springer, Berlin/Heidelberg, pp 249–257

Shuval HI, Yekutiel P, Fattal B (1985) Epidemiological evidence for helminth and cholera transmission by vegetables irrigated with wastewater. Jerusalem case study. Water Sci Technol 17(4/5):433–442

Shuval HI, Adin A, Fattal B, Rawitz E, Yekutiel P (1986) Wastewater irrigation in developing countries: health effects and technical solutions. Technical paper No 51. World Bank, Washington, DC

Tollner E (ed) (2007) Water profile of Bahrain. In: Cutler JC (ed) Encyclopedia of earth. Environmental Information Coalition, National Council for Science and the Environment, Washington, DC. FAO (Content Source). First published in the Encyclopedia of Earth December 4, 2007; Last revised date December 4, 2007; Retrieved January 26, 2013. http://www.eoearth.org/article/Water_profile_of_Bahrain

Chapter 7
Opportunities and Challenges of Using Treated Wastewater in Agriculture

Saif Ali Alkhamisi and Mushtaque Ahmed

Abstract The increase of the world population and the limitation of water in most of the countries led to over-pumping of traditional water resources especially the groundwater used in agriculture. Agriculture is using between 60 and 90 % of surface and groundwater. During the last two decades, the reuse of treated wastewater in agriculture had increased enormously due to the increase of food demand and expansion of agricultural areas. The shortage of water could be partially overcome by use of this new water source, Treated Wastewater (TWW). TWW can be used for irrigation under controlled conditions to minimize health risks arising from pathogenic and toxic pollution of the agricultural produce, soils, surface and groundwater. TWW reuse is one potential intervention strategy for developing nonconventional water resources. Reuse of TWW for irrigation and other purposes could contribute considerably to the reduction of 'water stress' and 'water scarcity' in the GCC countries. This chapter will highlight the factors that need to be considered when using TWW in agriculture especially the reuse in crop production, in addition to the main challenges and constraints facing the GCC countries in TWW reuse in agricultural production. It is a major challenge to optimize the benefits of TWW as a resource and to minimize the negative impacts of its use on human health. Different studies showed an increase of both the quantity and the quality of crops irrigated with TWW. In general, TWW has proved to be a very promising source of water for irrigation for crop cultivation. However,

S.A. Alkhamisi (✉)
Head of Field Crops Research, Ministry of Agriculture and Fisheries, Muscat, Oman
e-mail: saifalkhamisi@hotmail.com

M. Ahmed
Department of Soils, Water and Agricultural Engineering, College of Agricultural and Marine Sciences, Sultan Qaboos University, PO Box 34, Al-Khod 123, Oman
e-mail: ahmedm@squ.edu.om; ahmedm2316@gmail.com

S.A. Shahid and M. Ahmed (eds.), *Environmental Cost and Face of Agriculture in the Gulf Cooperation Council Countries: Fostering Agriculture in the Context of Climate Change*, DOI 10.1007/978-3-319-05768-2_7, © Gulf Research Centre Cambridge 2014

some constraints have been observed. These are related to economic, environmental and social issues. These constraints could be overcome by following certain practices and proper governmental policies and regulations.

Keywords Agriculture • Environmental constraints • Forage production • GCC • Treated waste water

7.1 Introduction

The world population has grown tremendously over the past 2,000 years. The latest official world population estimate, for mid-year 2011, is 6.9 billion. Populations require water for domestic and municipal usages; as an input in productive activities: agriculture, industry (including energy production) and services activities. Human beings need only about 5 L of water each day for cooking and drinking; according to WHO, however, good health and cleanliness require a total daily supply of about 30 L per person (11 m^3 $year^{-1}$) (UN 1994). Population growth is a direct determinant of increases in water demand for domestic uses. The demand of water for different uses indeed exceeds the supply in some areas but in most areas of the world the major cause of water scarcity is mismanagement towards available water resources.

Consequently, some countries have already reached the limits of their water resources.

The local population of the GCC countries adapted their lifestyles to the highly dry and arid climatic conditions that characterize this region. The population was small and was restricted to oases and upland areas endowed with higher rainfalls which would support cattle and crops. However, due to rapid population growth and the discovery and exploitation of oil resources starting from the 1960s, economic and social situations have drastically improved while imposing tremendous pressure on limited water resources (World Bank 2005). The world population is estimated to be 7 billion and expected to increase to more than 50 million by 2030.

Over the past three decades, the GCC countries have witnessed an unprecedented economic and social transformation. A significant portion of oil revenues has been used to modernize infrastructure and improve the living standards of the population. Water supply and sanitation services have been made accessible to a large percentage of the population. Life expectancy increased by about 10 years to 74 years during 1980–2000 and literacy rates increased from 20 % to about 80 % over the same period. Total water demand in all GCC countries has increased dramatically as a result of high population growth, improvements in the standard of living, industrial development in major urban centers and efforts to increase food self-sufficiency. According to World Bank (2005), the total water use for all sectors in the region increased by about four-and-a- half times from around 6 billion m^3 to 27 billion m^3 while the population more than doubled from around 14 million to

30 million during 1980–2000. The deficit has been and will be continuously met mainly by seawater desalination and mining of renewable and nonrenewable groundwater resources (World Bank 2005).

7.2 Agriculture and Treated Wastewater Reuse

7.2.1 Expansion of Agriculture

One of the primary reasons for the unsustainable exploitation of groundwater resources in some GCC countries has been the provision of direct and indirect subsidies to well excavation, pumps, fuel and other inputs as well as price support programs and trade protection. The governments intend to redistribute oil revenues for citizens, given that most of the employment in the agriculture sector is provided by expatriates, employment generation is not an objective of agricultural policy in GCC countries.

The increase of cropping areas in Gulf countries (Table 7.1) will consequently require sustainable conventional and non-conventional water resources (either surface or groundwater).

7.2.2 Treated Wastewater for Agriculture

Agriculture is the worlds' largest water user. However, due to population and urban growth, and scarcity of water, water resources are being transferred from agricultural users to municipal and industrial users, and are often accompanied by a

Table 7.1 The percentage of cultivated agricultural area to the area suitable for agriculture

Country	Total area	Area suitable for agriculture (Mha)	Actual cultivated agriculture area (Mha)	Percentage of agriculture area to area suitable for agriculture (%)
Kingdom of Saudi Arabia	2,250	52.0684	4.3595	8.37
Kingdom of Bahrain	0.074	0.0064	0.0035	54.94
United Arab Emirates	8.360	0.0742	0.0684	92.27
State of Kuwait	1.782	0.1538	0.0038	2.46
Sultanate of Oman	30.950	2.3000	0.0737	3.20
State of Qatar	1.150	0.6500	0.0123	1.89
Total	2,292	55.2528	4.5213	8.18

Source: Agriculture Development in GCC Countries (2011)

Table 7.2 Wastewater treatment capacity in the GCC countries[a]

Country	Existing capacity (m^3 day^{-1})	Additional capacity planned by 2015 (m^3 day^{-1})	Estimated cost of additional capacity ($Millions)[b]
Bahrain	221,000	280,000	493
Kuwait	697,000	795,000	1,399
Oman	106,000	230,000	405
Qatar	285,000	437,000	769
Saudi Arabia	1,952,000	2,224,000	3,914
UAE, of which:	965,000	1,607,000	2,828
Abu Dhabi	414,000	875,000	1,540
Dubai	260,000	400,000	704
Northern Emirates	291,000	332,000	584
Total	4,226,000	5,573,000	9,808

[a]Includes municipal projects only and not captive STPs serving real estate developments
[b]Cost is calculated on the basis of an average price of \$1,760 m^3 day^{-1}. This was the estimated cost of building STP capacity in Muscat, Oman in late 2008
Source: MEED (2010)

decrease in agricultural productivity. Therefore, the replacement of good-quality water resources with unconventional water sources, including TWW, as a new water resource for various uses has been seriously considered. This replacement can be a "win-win" solution to the regain and some of the water transferred from agriculture to municipalities and industry, can be put back to agriculture again (Ahmadi and Gary 2009). The reuse of TWW has been an important substitution for freshwater, especially for irrigation of agricultural lands, but there are many challenges such as pumping costs, health issues and environmental effects regarding the use of this resource.

In the GCC countries, more than one third of TWW is used to irrigate non-edible crops and fodder as well as for landscaping (Choukr-Allah 2010). The TWW has an important role to play in the GCC countries' water resources management. The present gap between water demands and available water resources has led these countries to consider domestic wastewater as an integral part of their water resources. As stated earlier presently, GCC countries recycle no more than 35 % of their total TWW, which contributes 2.2 % of their total water supply, being used mainly for landscaping, fodder crop irrigation, and some industrial uses. However, major plans for water recycling exist in most of these countries (Table 7.2). The main handicaps for reuse expansion are both social (psychological repugnance and religion) and technical (microbiological pollutants, potential heavy metals accumulation in irrigated soil, and industrial waste mixing). If only 50 % of domestic water supplies are treated and recycled in agriculture, recycled waters have the potential to meet more than 11 % of the GCC countries total water demands, and this could satisfy more than 14 % of their agricultural sector demands, and reduce fossil groundwater withdrawal by more than 15 % by the year 2020 (Al-Zubari 1997).

7.2.3 *Quality of Treated Wastewater*

7.2.3.1 Standards and Regulations

The reuses of TWW for irrigation have yielded positive economic, environmental and health restrictions and benefits but its reuse remain subject to physical and chemical properties and biological characteristics. The systems reuse this water according to the degree of treatment in three groups: tertiary wastewater treatment which is free from health hazards and compatible with the irrigation of all types of plants that are eaten raw as well as crops, orchards and pastures. While secondary wastewater treatment produces water for nursery flowers and palm trees, cotton, flax, fodder and vegetables (cooked crops) used in the food industry. This type of water can be for light and medium soil textures and prohibited for cattle rearing with milk or meat and plants cooked before eating. The primary treated water is not to be used by the non-timber trees, and to take all the environmental and health precautions, such as isolated farms, fences and the prohibition of entry for non-employees or where animals are raised at all, and the prevention of risks of dealing directly with the water.

International guidelines for use and quality standards of wastewater in agriculture exist to mitigate those impacts. The quality of TWW is typically defined in terms of its regulations set. These regulations define wastewater treatment levels and allow uses for the reclaimed water produced. When considered for use as irrigation water, the actual quality of treated wastewater ranges from totally unsuitable to ideally suitable, with most sources falling somewhere in between these extremes.

Most GCC countries have established conservatively low risk guidelines or standards (e.g. California standards) based on a high technology and high cost approach. However, high standards and high cost technologies do not always guarantee low risk because insufficient operational experience, costs, and maintenance costs, and regulatory control can have adverse effects (Choukr-Allah 2010).

Oman standards and regulations of wastewater reuse were established under Ministerial Resolution MD 145/93 which regulates TWW reuse and discharge. According to USEPA (2004) there are two main Omani rules, which regulate wastewater reuse:

1. Wastewater reuse, discharge and sludge disposal rules that include physico-chemical parameters such as suspended solids, conductivity, organic matters, and heavy metals, etc.
2. Wastewater standards related to biological characteristics.

Reuse regulations further classify wastewater use into two categories:

- Standard A – (200 FC/100 ml, less than 1 nematode ova L^{-1}) for irrigation of vegetables and fruit to be eaten raw, landscape areas with public access, controlled aquifer recharge, and spray irrigation
- Standard B – (1000 FC/100 ml, less than 1 nematode ova L^{-1}) for cooked vegetables, fodder, cereals, and areas with no public access (USEPA 2004).

Al-Zubari (1997) stated that GCC countries recycle no more than 43 % of their total TWW, which contribute to their total water supply, being used mainly in landscaping, fodder crop irrigation, and some industrial uses.

In the Kingdom of Saudi Arabia, in 2002 total TWW reached almost 548 $\times$ 10^6 million m^3, of which 123 million m^3 were reused. In 2003, 70 sewage treatment plants were in operation. The use of TWW is still limited (166 million m^3 in 2006), but it represents a potentially important source of water for irrigation and other uses (FAO-Water 2008). For safe reuse, standards have been established by different institutions to control the quality of the irrigation water. Current legislations in Saudi Arabia have been issued by Ministry of Water and Electricity (MWE 2006). Regulations identify two types of irrigations: restricted and unrestricted irrigations which depend mainly on kind of crops. Stringent regulations have been set to meet unrestricted irrigation which is intended for any crop and any type of soil without limitations (Alberta Environment 2000).

The regulations and standards for wastewater reuse in Kuwait were established by Environmental Public Authority (EPA) under Decree No (210) in 2001 which stated the criteria for TWW reuse in terms of maximum allowable (WHO 2006) limits.

Abu Dhabi emirate has its own guidelines for wastewater and biosolids called Recycled Water and Biosolids Regulations 2010 issued by Regulation and Supervision Bureau through Water, Wastewater and Electricity sector of the Emirate of Abu Dhabi. These Regulations aimed to provide a clear legal framework for managing Recycled Water and Biosolids. They ensure the safe, environmentally beneficial and economic management of Recycled Water and Biosolids by Disposal Licensees (RSB 2010)

7.3 Challenges of Treated Wastewater Reuse

7.3.1 Costs Associated with Treated Wastewater Reuse

The TWW can contribute not only in decreasing the water scarcity, but also to increase social and economic development. Thus the reuse of TWW could be important means to sustainable development. Several studies claim that the optimal wastewater treatment level is affected by costs, hazards and benefits. So, lowering the wastewater treatment level decreases fertilization costs because of the increased levels of available nutrients left in the water, and irrigation costs decrease if water prices reflect the lower treatment costs. Agricultural yields and prices may decrease according to differences between levels of nutrients needed by crops and those available in wastewater.

When considering the reuse of TWW for irrigation, factors of the costs-benefits should be considered. Some studies related to economic factors with the social acceptance of TWW are reported here. Al-Dadah (2003) differentiates the financial and economic dimensions of TWW reuse. He uses average incremental cost (AIC) per cubic meter of TWW as the bases for the financial analysis. It is also suggested to arrange the water and TWW prices by considering respectively water and TWW charges based on the AIC to recover the operational and replacement costs. It is suggested that the evaluation of TWW irrigation techniques should consider cost of treatment, cost of irrigation, level wastewater treatment required, water use efficiency, health risk and the cost of distribution (Özerol and Günther 2005). Papadopoulos (1995) argues that the economical evaluation of irrigation with TWW possesses two major difficulties. The first difficulty is the valuation of non-financial aspects like reduction of environmental pollution or health risks, and the second one is the allocation of treatment plant costs between the producer and the user of TWW. He argued that if the above difficulties are handled, the cost-benefit analysis of TWW reuse in irrigation can be made based on the following elements:

- Estimation of least-cost disposal options that meet environmental and health standards
- Identification of the demand areas for wastewater and the corresponding cost of transportation
- Incremental treatment cost of wastewater
- Price of treated wastewater

Several treatment plants are currently under construction in the different GCC countries. The annual TWW production increased from 1.161 billion m^3 in 2000 to be 1.57 billion m^3 in 2010. As many rural areas in the GCC countries are not connected to sewerage systems, the wastewater that is available for treatment constitutes about 25 % only of domestic and industrial demands (Al-Rashed and Sherif 2000).

7.3.2 *Economic, Social and Environmental Constraints*

Although reuse of wastewater has a high positive potential to environmental relief and social economic development, obviously there is also the danger of the opposite effect if the reuse schemes are not properly planned and managed. For instance, as a primary disadvantage, the demand for wastewater is usually only during the growing season whereas its production is continuous, which might cause high environmental and health hazard risk if the water is not treated and stored adequately (Kretschmer et al. 2002). Therefore the treatment and storage of wastewater should be made accordingly to prevent both hazardous cases and high costs of storage. The acceptance of TWW has started to increase among the farmers in most of GCC countries. The farmers in other Arab countries like Jordan, Tunisia and Syria are using TWW for irrigation for almost all food crops. The studies showed significant benefits from using TWW and low impacts on soil and crops from such irrigation, still the challenges are present concerning its utilization. The challenges could form barriers in using TWW for crops irrigation. The constraints towards maximum use of TWW can be classified as economic, social and environmental. The cost of water has to be acceptable to farmers. If farmers find free source of water they will not go for TWW. The transportation cost of TWW from Sewage Treatment Plants to agricultural areas could be another economical constraint, since most of big STPs are located in the cities where the concentration of people is dense. From the environmental aspect there are potentially positive and negative impacts that should be considered. Presence of pathogens in TWW, chemical contaminants or heavy metals because of insufficient treatment could form a critical constraint towards efficient reuse in crop production. The major environmental risks from TWW occur through the contamination of water with chemical substances and pathogens, due to poor treatment and/or inadequate management guidelines. The level of impact depends on the degree of purification, the method and the location of reuse and it can be observed in the form of pollution in soil, groundwater or surface water (Papadopoulos 1995; Kretschmer et al. 2002). It may cause also soil quality problems in the long run such as accumulation of salts and heavy metals in the soil. Some substances that can be present in wastewater in such concentrations that they are toxic for plants or lead to environmental damage and potential harm to the soil due to heavy metal accumulation and acidification could be classified as disadvantages. Those disadvantages could lead to the risk of potential harm to groundwater due to heavy metals, nitrate and organic matter, and human health by spreading pathogenic germs (Kretschmer et al. 2002).

7.3.3 Risks During Transportation of Treated Wastewater from STPs to Agricultural Areas

The transportation of untreated and TWW from sewage treatment plants (STPs) involves slight technical risks related to, for example, pumping station failures or tunnel cave-ins. These risks and their effects increase as the length of transfer lines and the number of pumping stations increase. Most of the STPs are located in the cities where the agricultural areas are far enough to transport TWW to those areas. It is practical to optimize building the new STPs near the current agricultural areas or far. The further away from agricultural areas is positioned, the greater the overall costs and the operating costs related to pumping. In addition, it increases the amount of excavation and the environmental impacts. Each kilometer further away will increase the costs. From the other side, building STPs far away from residential domestic areas will increase the cost of transferring the untreated wastewater to those STPs.

Large STPs usually achieve better treatment results than small ones and also have higher operational reliability. The long transfer lines inevitably related to large plants carry their own risks. With regard to overall risks, it does not make a significant difference whether wastewater is treated in one large plant or several small ones, but the controllability of a large plant is usually better. A good example is Sulaibiya STP in Kuwait which contributed in reducing the annual demand from non-potable sources. The TWW is transported from Kuwait city to Al-Abdali, which is about 120 km far where the TWW is used for agricultural crops cultivation.

In Oman, the biggest STP is Al-Ansab, which is producing 14,415 m^3 per day (as of 2010). It is located in Muscat, about 100 km from Al-Batinah, where extensive agriculture is found in the region. Consequently, along those distances, the TWW will have to be transferred through pipes and may have to be stored between places to place. That storage has the risk to cause increase of pathogenic microorganisms.

7.3.4 Farmers' Acceptance of Treated Wastewater Reuse

Social issues play a significant role in water reuse initiatives and should be addressed with adequate political will accompanied by awareness programs to overcome cultural, religious and social objections. Acceptance of farmers, retailers and consumers is the most sensitive and important issue. Farmers are not going to reuse TWW, if their product cannot be sold. Consumers will not buy products where TWW was used unless it is proven to be safe. The above mentioned constraints could be mitigated or overcome by certain specific practices. Treated wastewater must only be reused for the uses for which permit was issued. Quality monitoring and process controls should be supported and strictly applied. When TWW quality does not meet the fixed standards and regulations then reuse must cease. Surface or subsurface (not sprinkler) irrigation should take place in all type of crops cultivation.

Kuwait has a national sewage network which is estimated at 800 linear km. The sewage wastewater is transmitted from 98 % of the facilities in the country. The Sulaibiya Wastewater Treatment and Reclamation Plant is considered by far the largest facility of its kind in the world to use reverse osmosis (RO) and ultrafiltration (UF) membrane-based water purification systems. The plant's daily capacity is 375,000 m^3, which could be expanded to 600,000 m^3 day^{-1} in the future. It is believed that TWW will ultimately contribute to 26 % of Kuwait's overall water demand (Choukr-Allah 2011). Produced TWW in Kuwait is transported from the WWTPs to reuse places through underground distribution networks or tanker truck deliveries to the main farming areas of Sulaibiya, Abdali and Wafra areas. TWW is stored and disinfected (chlorinated) before being distributed to points of utilizations. Six effluent storage reservoirs about 30 km away from the center of Kuwait City of total storage capacity of 38,000 m^3 day^{-1} are now increasing to 68,000 m^3 day^{-1} through building new reservoirs (Al-Anzi et al. 2012)

7.4 Benefits of Treated Wastewater Reuse

When considering wastewater reuse for irrigation, an evaluation of the advantages, disadvantages and possible risks has to be made. The advantages include improvement of the economic efficiency of investments in wastewater disposal and irrigation, conservation of freshwater sources and use of the nutrients of the wastewater (e.g., nitrogen and phosphate). Wastewater is normally produced continuously throughout the year, whereas irrigation is mostly limited to the growing season. The GCC countries follow similar methods in the disposal of TWW. A high percentage of TWW is reused in irrigation of forage agriculture land or in landscaping while the remaining is disposed into the sea after advanced treatment techniques (Choukr-Allah 2011). For instance, experiments have shown that the use of TWW in Saudi Arabia as a supplemental irrigation has not only increased crop production, water use and nitrogen use efficiencies but also served as a source of plant nutrients. The use of TWW saved up to 50 % application of inorganic nitrogen fertilizer if TWW contains 40 mg N per liter (Hussain and A-Saati 1999).

7.5 Case Studies from Oman

7.5.1 Utilization of TWW in Forage Production

The Ministry of Agriculture and Fisheries of Oman represented by the Directorate General of Agriculture and Livestock Research, has been promoting the use of TWW for agriculture through research and studies on different crops, especially forage crops. It has implemented a series of experiments for the use of tertiary treated wastewater on forage crops like barley, maize and sorghum in different seasons (winter and summer) as well as analysis of soil and irrigation treated wastewater for pH, salinity and micro-elements or heavy elements to find out their suitability for animal feeding which will be reflected on human nutrition and health.

To ascertain these results and the awareness of the farmers, the Directorate General of Agriculture & Livestock Research initiated the joint 'Pilot Project' to explore the possibilities of tertiary treated wastewater utilization at Saham Sewage Treatment Plant (STP) for the cultivation of selected fodder crops, a unique step in reducing the pressure on groundwater resources and trying to bridge the gap on water scarcity. The objectives of the project were mainly to take advantage of the quantities of TWW and areas for cultivation and production of seasonal forage crops. Other objectives included evaluation of forage crops for winter and summer cultivation (sorghum, maize, barley); to study the impact of tertiary treated wastewater on the soil at the targeted sites and to evaluate the projects' techno-economic parameters for future planning and utilization of TWW.

The farm site located adjacent to Sewage Treatment plant (STP) of Saham was selected for conducting experimental activities of the project. At the Saham farm site, entire fencing was undertaken as well as civil and mechanical works related to pumping and drip irrigation system. The total area of the site selected was 8 Feddans (1 Feddan = 4,200 m^2) of which the experimental area was 6 Feddans. The production of treated effluent of Saham STP was estimated to be 900–1,200 m^3 day^{-1}. Three cross-sections were selected for soil sampling at the site to examine soil suitability for cultivation. Three crops namely Barley (*Hordeum vulagare* L.), sorghum (*Sorghum bicolor* L.) and maize (*Zea mays* L.) were selected for growing in crop-rotation as fodder crops at appropriate layouts. The economic analysis was carried out on the basis of information made available on input costs, output yield per crop per season and response on sale prices consulted through main fodder markets. With certain assumptions, the economic indicators were worked out for testing the feasibility/viability of the project and on this basis it was safe to conclude that the project indicated strong economic viability. The practices followed for cultivation, scale of operation and farm management were the critical parameters for obtaining successful results in considering the commercial application of such experimentation.

From the above results it can concluded that the forage yield of sorghum and maize was increased by 30 % under TWW irrigation in comparison with that using fresh water whereas this increase was 43 % in barley crop. The contents of toxic elements in the plant tissues of the fodder produced using TWW were found to be below the standard/ideal safe limits recommended. The encouraging results of the pilot project directly provided prudent guidelines for sustainable growth of agriculture on such TWW utilization for the selected crops namely barley, sorghum and maize crops in rotation whose economic analysis showed the positive results indicating the economic viability of the project with average IRR (Internal Rate of Return) as 26 %. An average payback period as 2.6 years endorsing the economic viability of the project. For commercial application and private sector investment, these results were likely to be enhanced. At the same time organizational efforts and management cost might also increase. There is rich potential to use TWW for fodder crop cultivation and this seemed as economically viable project to attract investment from the private sector. The cultivation of fodder by using TWW utilization is likely to have favorable environmental impact on green land development. The economic and social impact is also likely to be the most favorable due to better crop yields and better economic returns realized as per the economic analysis. Some government support would be essential to promote such cultivation and supply of TWW. In terms of "food security" which is top on the agenda, promotion of such projects utilizing the TWW for increasing fodder supply is of paramount importance on the national level (MAF 2011).

7.5.2 *Yield and Water Use Efficiency of Forage Maize (*Zea mays* L.) Under Treated Wastewater Irrigation*

Field studies were conducted during the 2006/07 season to determine the effect of water quality (tertiary treated wastewater and fresh water), water quantity (1.4ETc,

1.0ETc and 0.6ETc), and their interaction on the growth, yield and water use efficiency of forage maize. In addition, the chemical composition of forage maize plant that had been irrigated with TWW was evaluated in comparison with those irrigated with fresh water. Soil moisture distributions and salinity redistributions were monitored throughout the experiments.

The results indicated that TWW leached more salts down the profile than freshwater. The TWW also reduced the SAR (Sodium Adsorption Ratio) by 74 % whereas freshwater by 68 %, but freshwater had a higher SAR to start with. The higher the quantity of water applied, the higher was the salinity and SAR reductions. Freshwater treatments were observed to have higher moisture content in comparison with the TWW. This was attributed to the fact that TWW contained dissolved organic matter that slightly improved the physical conditions of the soil which resulted in increased water penetrations, and contained higher amounts of nutrients that resulted in vigorous plants which abstracted more water. The results also indicated that plants irrigated with TWW contained higher nitrogen concentrations at all levels of water quantities than those irrigated with freshwater. But K, P, Ca, Mg, Fe, S, B, Zn, Cu and Mn uptakes of forage maize did not show any significant differences between fresh and TWW. Plants irrigated with TWW had higher growth rates (in term of plant height) in comparison with those irrigated with freshwater. The number of leaves/plant, leaf length and leaf area (cm^2) did not show any significant differences among water types, water quantities or their interactions. The TWW had shorter time for 50 % male and female flowering of forage maize plants than freshwater, indicating earlier maturity. Plants irrigated with TWW had higher chlorophyll content than those irrigated with freshwater for all levels of water applications. The TWW gave higher average green forage yields (60.79 t ha^{-1}) and dry matter yields (11.57 t ha^{-1}) than fresh water which yielded 36.27 and 9.46 t ha^{-1} of green forage and dry matter respectively. Plants irrigated with TWW were more efficient in using the water than those irrigated with freshwater for all water quantities. The highest water use efficiency (3.51 kg m^{-3} DM) was achieved with TWW under the 1.0ETo water applications. These studies concluded that TWW irrigations increased yields of forage maize and their water use efficiency without significantly affecting any metal accumulations in the soil or plant leaves (Alkhamisi et al. 2011).

7.6 Conjunctive Use of Reclaimed Water and Groundwater in Crop Rotations

With the increasing scarcity of freshwater available to agriculture, the need to use TWW in agriculture has increased. Currently most of the Gulf countries, including Oman uses TWW to irrigate public gardens and green strips in urban area. Irrigation demands of these amenities vary during the year, while the supply of reclaimed water from sewage treatment plants (STP) remains reasonably steady. The surplus

TWW can be used to grow seasonal crops, or stored for future in aquifer recharge, or disposed in the sea. Often, water quality requirements for aquifer storage are very stringent, and meeting these requirements is costly. Therefore there is a need to maximize the use of TWW, by growing short season crops throughout the year, changing the area under cultivation of such crops, and supplement TWW with groundwater. By doing so, TWW will not be disposed to the sea, it need not to be injected to the aquifer, stress on groundwater will be minimal, while crop production will be maximized. In order to better utilize the TWW, the Oman Wastewater Services Company (OWSC) is planning to pipe it to the AlBatinah region of Oman. The area is of vital importance to the agricultural economy of the Sultanate of Oman, but 52 % of the area is abandoned due to soil and groundwater salinity. In this study, we explored how TW from an STP can be used directly, without Aquifer Storage and Recovery, as a source of irrigation water in conjunction with groundwater for agriculture. Average data from Muscat, Oman in the years from 1996 to 2010 was used for calculation of crop water requirement. Wheat, cowpea and maize were chosen as crops to be grown in rotation through the year.

Results show that by using TWW conjunctively with groundwater cropping area can be increased from 695 to 2,245 ha (323 % increase) of wheat, 313 to 782 ha (250 % increase) of cowpea and 346 to 754 ha (318 % increase) of maize. Of the total irrigation requirement 24.24 mM^3, 57.6 % was met with TWW and 42.4 % was met with groundwater. Therefore, the planners should consider piping TWW to areas where groundwater of good quality is available to conjunctively use and meet crop water requirements, than piping it to areas where groundwater is saline and unsuitable for irrigation. This will prevent disposal of TWW to the sea and minimize stress on fresh groundwater zones.

7.7 Conclusion

Water is the major challenge facing most of the arid and semi-arid countries including the GCC countries. The TWW has proved to be a very promising source of irrigation water for crop cultivation. The reuse of TWW targets agriculture predominantly especially in GCC countries for irrigation and landscaping. There are economic, institutional, health, and environmental constraints that hamper the sustainable and safe reuse in agriculture. To address these limitations concerted effort and commitment is required to boost the volumes of TWW produced. For maximum utilization of this resource, the cost of water has to be acceptable to farmers. If farmers find free source of water they will not go for TWW. The other challenge is the transportation cost of TWW from Sewage Treatment Plants to agricultural areas, since most of big STPs are located in the cities where the concentration of people is dense. Social issues play a significant role in water reuse initiatives and should be addressed with adequate political will accompanied by awareness programs to overcome cultural, religious and social objections. Acceptance of farmers, retailers and consumers is the most sensitive and important issue. Farmers are not going to reuse water, if their product cannot be sold. Consumers will not buy products where TWW was used unless it is proven to be safe. International guidelines for use and quality standards of wastewater in agriculture exist to mitigate those impacts. Several studies were conducted in Oman and recommended the reuse of TWW in agriculture. They highlight the impacts of TWW on agricultural production. However, some constraints have been observed. These are related to economic, environmental and social issues. These constraints could be overcome by following certain practices and proper governmental policies and regulations.

References

Ahmadi L, Gary PM (2009) Planning and management modeling for treated wastewater usage. Irrig Drain Syst 23(2–3):97–107

Al-Anzi B, Abusam A, Shahalam A (2012) Assessment of wastewater reuse in Kuwait and its impact on amounts of pollutants discharged into the sea. J Environ Prot 3(8A):935–939

Al-Dadah JY (2003) Socioeconomic aspects of wastewater reuse in Gaza Strip. In: International seminar on wastewater reclamation and reuse, Izmir, Turkey

Al-Rashed MF, Sherif MM (2000) Water resources in the GCC countries: an overview. Water Res Manage 14:59–75

Al-Zubari WK (1997) Toward the establishment of a total water cycle management and re-use program in the GCC Countries. In: Gulf water conference (1997) No 3, Muscat, Oman (08/03/1997) 120(1–2): (171 p.) (23 ref.), [Notes: Selected papers], pp 3–14

Alberta Environment (2000) Guidelines for municipal wastewater irrigation. http://environment.alberta.ca/. Accessed 20 Mar 2012

Alkhamisi SA, Abdelrahman HA, Ahmed M, Goosen MFA (2011) Assessment of reclaimed water irrigation on growth, yield, and water-use efficiency of forage crops. Appl Water Sci J 1:57–65

Choukr-Allah R (2010) Wastewater treatment and reuse. In: El-Ashry M, Saab N, Zeitoon B (eds) Arab environment: water: sustainable management of a scarce resource. Report of the Arab Forum for Environment and Development (AFED), Beirut, Lebanon, pp 107–124

Choukr-Allah R (2011) Wastewater treatment and reuse in the Arab Regions. Biosalin News 12 (2):4–5

FAO-Water (2008) AQUASTAT, Land & Water (2012) FAO's information system on water and agriculture. http://www.fao.org/nr/water/aquastat/countries_regions/saudi_arabia/index.stm. Accessed 21 Mar 2012

General Secretariat of GCC (2011) Agriculture development in GCC countries, 5th edn. Economics Affairs, Riyadh, KSA

Hussain G, A-Saati AJ (1999) Wastewater quality and its reuse in agriculture in Saudi Arabia. Desalin J 123:241–251. http://linkinghub.elsevier.com/retrieve/pii/S0011916499000764. Accessed 23 Feb 2012

MEED Insight (2010) An overview of the GCC wastewater and water. http://www.meed.com/research/gcc-wastewater-2009-the-drive-for-capacity/2025082.article. Accessed 14 June 2012

Kretschmer N, Ribbe L, Gaese H (2002) Wastewater reuse in agriculture, technology resource management and development. Special Issue Water Manage 2:37–64

MAF (Ministry of Agriculture and Fisheries) (2011) Final report – the pilot project on utilization of treated wastewater in forage production. Directorate General of Agriculture and Livestock Research. Ministry of Agriculture and Fisheries. Sultanate of Oman. 20 English + 20 Arabic pages

Ministry of Water and Electricity (MWE) (2006) Guidelines for municipal wastewater reuse for irrigation, controls – conditions – offenses and penalties (Arabic). Kingdom of Saudi Arabia. http://www.moa.gov.sa/cs/allContent/EL_MINISTRY/raae/raae1.htm

Özerol G, Günther D (2005) The role of socio-economic indicators for the assessment of wastewater reuse in the Mediterranean region. In: Hamdy A, El Gamal F, Lamaddalena N, Bogliotti C, Guelloubi R (eds) Non-conventional water use: WASAMED project Bari: CIHEAM-IAMB 2005, pp 169–178. 3. WASAMED (Water Saving in Mediterranean Agriculture) workshop, 07-10/12/2004, Cairo (Egypt). http://ressources.ciheam.org/om/pdf/b53/00800760.pdf. Accessed 11 Mar 2012

Papadopoulos I (1995) Non conventional water resources: present situation and perspective use for irrigation. In: International seminar on economic aspects of water management in the Mediterranean Area. Proc seminar held in Marrakech, Morocco, 17–19 May 1995. CIHEAM, pp 54–76

RSB (2010) The recycled water and biosolids regulations. Issued by the Regulation and Supervision Bureau for the water, wastewater and electricity sector in the Emirate of Abu Dhabi. Consultation version 0.2 www.rsb.gov.ae Page 1 of 34

UN (1994) Population and water resources (contribution by FAO). United Nations Population Information Network (POPIN), UN Population Division, Department of Economic and Social Affairs, with support from the UN Population Fund (UNFPA) (1994). http://www.un.org/popin/fao/water.html. Accessed 4 Mar 2012

United States Environmental Protection Agency (USEPA) (2004) Guidelines for water reuse-chapter 8, EPA/625/R-04/0. http://www.epa.gov/

WHO, World Health Organization (2006) Compendium of standards for wastewater reuse in the Eastern Mediterranean Region. Regional Office for the Eastern Mediterranean, Regional Center for Environmental Health Activities, CEHA, Cairo, Egypt

World Bank (2005) A water sector assessment report on the countries of the Cooperation Council of the Arab States of the Gulf. Water, Environment, Social and Rural Development Department. Middle East and North Africa Region. Report No. 32539-MNA, 31 Mar 2005

Chapter 8
Investments in Foreign Agriculture as a Gulf State Food Security Strategy: *Towards Better Policy*

Benjamin Shepherd

Abstract Large-scale investments in foreign agricultural land are becoming an increasingly important part of the food security policy mix for some Gulf States (GS). The investment projects are usually made by corporate enterprises but supported diplomatically, and often financially, by GS governments. Fieldwork by the author in Cambodia, the Philippines and Ethiopia has produced evidence that land-leasing investment projects are proving to be a far from ideal mechanism for increasing "food security." Few projects in these countries are showing signs of providing reliable, long-term supplies of food staples for export back to the investing country. Further, if food security is understood in terms of reducing hunger, some projects are actually increasing food insecurity: they are dispossessing and impoverishing local communities, even in areas already affected by poverty and hunger. Governments in some of these host countries appear to find eviction orders against their own people easier to enforce than proper regulation of foreign investors' activities. Forcible evictions of the rural poor without adequate compensation are bad for the affected communities; they also add risk to the investment environment. This chapter argues that GS policy-makers who are facilitating foreign agricultural investments as part of their food security policy mix need to regulate their investor-enterprises much more closely if they wish to see successful "food security" outcomes. One important strategy explored is the opportunity to invest in the host countries' farmers instead of in their farmlands. Instead of projects which remove the productive capital of the land from access by the rural poor, investments can be structured to leverage the capacity of rural farmers to increase their production yields. By providing these farmers with the financial and technical support to achieve greater productivity and by granting them access to the important GS markets for their surpluses, there is the possibility to deliver positive food security outcomes. These can be measured both in terms of

---

B. Shepherd (✉)
Centre for International Security Studies, University of Sydney, Sydney, NSW, Australia
e-mail: benshepherd@mac.com

S.A. Shahid and M. Ahmed (eds.), *Environmental Cost and Face of Agriculture in the Gulf Cooperation Council Countries: Fostering Agriculture in the Context of Climate Change*, DOI 10.1007/978-3-319-05768-2_8,

food availability and reliable sources of staple food supplies and in terms of reducing poverty and hunger in the developing country.

Keywords Cambodia • Ethiopia • Food security • Foreign agriculture • Gulf States • Philippines • Vietnam

8.1 Introduction

A number of the Arab states of the Arabian Gulf have become involved in land leases, sometimes referred to as "land grabs," that acquire usage rights to large areas of agricultural land in developing countries (Alshareef 2009; Anseeuw et al. 2012). The ostensible purpose of these investments is to develop agricultural productivity in pursuit of "food security" (England and Blas 2009; Kingdom of Saudi Arabia Ministry of Foreign Affairs 2010; Reuters 2008; Kuwait Finance Minister al-Shamali cited in Reuters 2009; Walid 2009). While projects are usually executed by corporate enterprises, the states are involved in the negotiation of arrangements with developing country governments, the establishment of some large investment projects and the provision of financial and other support to enterprises engaging in such projects. As such, these activities form part of the broader food security policy agendas being pursued by Gulf States (GS) in the light of limitations on domestic agriculture and perceived needs to secure future food supplies (Shepherd 2010). Opponents of such projects, however, have been vocal in their criticism, arguing that they are likely to deprive and harm the communities who depend for survival on the land being granted to investors (Blas 2008; Grain 2009; Monbiot 2008). Despite a widespread legitimizing discourse to the contrary that projects will be "win-win"—offering positive development outcomes for the developing country as well as profit and exportable food supplies for the investor (Bladd 2010)—evidence indicates that there is some basis to the criticisms made (Shepherd 2012a, b). Elsewhere, my colleagues and I have argued that projects are unlikely to deliver the desired food security outcomes for GSs because of the hostile investment environment created by deals which dispossess local communities of the land they depend upon and set investors in conflict with those communities (Tetreault et al. 2012). These pathologies have been observed first-hand during field research in Cambodia and Ethiopia (Shepherd 2012a, b). In the former, violent eviction of farmers to make way for investors is not uncommon while in the latter; communities are sometimes pre-emptively evicted so that the government can claim lands granted to foreigners are "unused." Many of these projects are still in the early stages, nevertheless there is a dearth of success so far and prospects for large-scale production to supply Gulf markets appear unlikely (Woertz 2011). This is problematic for the GS governments if they are indeed expecting to see a consistent inflow of staple foodstuffs from foreign agriculture projects, and it is problematic for the local communities who are suffering from the establishment of the projects even though major productivity gains—and employment and other possible benefits from the projects—are a long way from becoming reality.

The purpose of this chapter is not to repeat the negative arguments against land deal investments which have been thoroughly articulated elsewhere (Borras et al. 2011; Daniel and Mittal 2009; Rahmato 2011; Shepherd 2012c; Speildoch and Murphy 2009; World Bank 2010). Instead, the objective of this chapter is to make the positive case for an alternative policy approach which is geared towards achieving the desired food supply/food availability outcomes for GS investors but which seeks to mitigate the pathologies evident from the land-centric investment model. The argument put forward is that a policy framework which facilitates local communities to produce surpluses for export will be more successful than acquiring land-use rights which exclude those local communities from the land they rely upon to survive. The experience of agricultural development in Vietnam is used as an example. Lessons from Vietnam can be used to help inform better policy-making by GSs seeking to build reliable sources of staple food supplies by investing in the development of the agricultural sectors of least-developed countries (LDCs). This chapter canvasses some possible approaches for facilitating this.

The first part of the chapter establishes the argument for GS governments to re-think their policies in support of foreign agriculture, using the evidence from the field in Cambodia and Ethiopia. These two have been selected because they are important targets of land deals by GCC investors and are countries that present considerable opportunity—and a dire need—for improving agricultural productivity, given suitable investment. They were also the subject of prior field study by the author. The second part of the chapter makes the case for a policy framework that focuses on developing the capacity of local farming communities in the developing country host, using data gathered from interviews with policy -makers, -analysts and -practitioners in Vietnam to make the argument. Vietnam has been selected because it is a case that demonstrates the potential for success of alternatives to land-centric investments. The chapter concludes with some suggestions for how such a policy framework might look and with some recommendations for continuing research to validate and operationalize such a model for foreign investment in LDC agriculture.

8.2 Conceptualising the Problem

It is problematic that food security is often thought about in ways that focus on "securing" food producing resources and having "secure" food supplies as distinct from thinking about securing humans against hunger. This is particularly troublesome given the high levels of agricultural productivity in the world, but the high levels of hunger which are a product of the failure of global society to share that production equitably (Sachs 2006; Sen 1981, 1990; Sen and Dreze 1987; Uvin 1994) and an incredible wastage of food (Gustavsson et al. 2011). The GSs' pursuit of usage rights over agricultural land in developing countries is an example of a "resource-securing" approach to food security and reflects the hegemony of this way of thinking about food security. That these projects can lead to worsening hunger—by depriving communities of their subsistence and sources of income—not to mention increasing the possibility of physical violence in some situations (Shepherd 2012a, b), speaks to the limitations of approaching "food security" in this way. Elsewhere I have laid out an argument for rethinking food security more holistically, especially to consider—and to seek to address—the problem of hunger in the world (Shepherd 2012d). The following discussion recognizes the dominance of the current way of thinking about food security, including amongst GS policy-makers. However, the policy approach suggested herein aims to be compatible with a framing of food security that privileges the securing of vulnerable populations from the risk of hunger. Moreover, the policy framework explored here proactively seeks to address food security in terms of tackling the problem of hunger. At the same time, it is argued that the quest for reliable long-term sources of food staples by the GSs need not be incompatible with that goal. Instead, as this chapter will make clear, the opening of GS markets to the agricultural produce of LDC farmers actually helps develop food security by improving the farmers' resilience to risks of hunger as well as bolstering food supply availability for the GSs.

Pursuing food security by acquiring usage rights to agricultural land in LDCs is predicated on a questionable assumption that there is much agricultural land in least-developed countries that is unutilized and available for exploitation. In Cambodia and Ethiopia this is not the case. While much agricultural land is *under*-utilised in terms of low per-hectare productivity, even in remote districts, agricultural land is nearly always depended upon by some—frequently very poor—communities for survival. Even in apparently unsettled regions in Ethiopia, land plays an important role in supporting mobile pastoralists during their annual movements. While in Cambodia's forest-land areas, which are targeted by the government for—frequently illegal (Global Witness 2007)—logging and turning into "agricultural land," the land is crucial for indigenous communities who do not farm in the traditional sense but are nevertheless dependent upon its natural productivity for their food and income. Others point out that this assumption in the availability of unutilized land is false for many other LDCs as well (Cotula et al. 2009). On the one hand, this situation suggests that in many circumstances where land is enclosed for use by foreign investors there will be communities who are excluded from the land and who are likely to suffer consequences as a result. On the other, however, the *under*-utilisation of agricultural land speaks to the opportunity to increase food production by increasing the productivity of those lands.

Considerable opportunity exists for the improvement of agricultural productivity in both Cambodia and Ethiopia, despite their widely different agricultural sectors, policy regimes and natural resource endowments. Both countries, like many other LDCs, are heavily dependent on agriculture with large proportions of the population reliant upon it for survival; Ethiopia 84 % and Cambodia 72 % (Federal Democratic Republic of Ethiopia [FDRE] 2010; National Committee for Population and Development 2010). This in turn suggests that these farming communities offer a considerable opportunity for GS investment. By re-thinking food security in terms of securing the hungry and vulnerable instead of as zero-sum competition over scarce resources, the host countries' agrarian poor become a higher priority than the corporations and investment vehicles of the investor country. From this perspective, an investment strategy that seeks consistent food surpluses for export would focus on investing in the smallholder agricultural communities instead of trying to take control over the land they rely on for survival. The beauty of such a strategy is that it avoids the large-scale dispossession and impoverishment of rural communities that are a consequence of evicting them from their land—and which create hostile environments for investors—and instead leverages the capacity and resources of those communities to create surplus production. Although it might sound utopian or idealistic, such strategies have the ability to address food security both in terms of increasing overall food availability and protecting the vulnerable against threats of impoverishment and hunger. And thus, as a long-term food security strategy for both increasing the resilience of the agrarian communities and for creating food surpluses for export, such a strategy has a higher likelihood of success than land acquisitions. This is an argument for investors—and more importantly for the GS policy-makers with power to regulate the behaviour of the enterprises involved—to invest in *people* and their capacity to become successful partners for mutual long-term benefit. As suggested by a Vietnamese policy-maker, it could be termed an *investment partnership* approach.[1]

8.3 An Investment Partnership Approach

In putting forward the argument for an investment strategy that focuses on smallholder agriculturalists and their communities instead of focusing on acquiring land, I make two assumptions. One is that the pursuit of an investor's commercial objectives—including agricultural productivity improvements and production of surpluses for export (and indeed for profit) need not be in contradiction with the provision of protection to the poor, hungry and vulnerable from hunger and other human insecurity. The other is that commercial interests will tend to look after themselves: The purpose of commercial investment is to make a profit (even if financially supported by the investor state governments to get projects off the ground) and business enterprises will strive to maximise their profit within the constraints placed upon them. Prioritising the needs of the poor and hungry over the desires of profit-making enterprises, and placing certain constraints on commercial activity towards that end, does not remove the profitable opportunities from play nor does it prevent investors and corporations from making profits. If an enterprise believes that the opportunities become unattractive as a result of constraints placed upon them, there is no obligation upon them to continue to play and other more efficient enterprises can take the opportunity to participate.

[1] Term suggested by a Vietnamese policy maker interviewed in Hanoi, 8 February 2012.

It should also be made clear that the decision to focus on the GS governments as agents for affecting change to current practices of foreign agricultural investment projects in LDCs has been made for three important reasons. First, because offering policy advice to the regimes of the LDC states hosting such investments is unlikely to affect change. This is partly because—even with the best of intentions—these governments frequently do not have the enforcement capacity to deliver on sound policy: well-designed laws that are not implemented with any good effect are evidence of this in both Cambodia and Ethiopia. It is also partly because LDC regimes frequently work to a security praxis that seeks to maximise elite gains while balancing the risk of local domestic instability (Ayoob 1995; Bueno des Mesquita et al. 2005). Especially in a kleptocratic system like Cambodia's, unless policy prescriptions can offer improved benefits to the regime to offset greater human rights protections, there is little likelihood of serious engagement in doing things differently.

The second reason is because multilateral institutions and the international organizations that play an important role in tackling "food security" as an international matter—such as the FAO and the World Bank—have demonstrated themselves to be incapable of affecting positive change in this area. They support the idea of a Code of Conduct to manage foreign land acquisitions which has no enforceability and, which I have argued elsewhere, actually perpetuates the resource-securing approach to food security instead of addressing the needs of the poor and vulnerable (Shepherd 2012c).

The third reason, conversely, is that the investor state governments *do* offer a strong opportunity to influence change. Saudi Arabia, by way of example, claims to be seeking particular food availability objectives through its establishment of the King Abdullah Initiative for Saudi Agricultural Investment Abroad (KAISIA),[2] but it is clear from the evidence of the previous studies that these objectives are unlikely to be met by the current, flawed, strategies focussed on land acquisition. The GSs in general are investing heavily, both in financial and diplomatic terms, in supporting their enterprises making investments in LDC agriculture. If it is true that these states are seeking long-term food availability benefits—and are not, as might be a cynical interpretation of the Saudi situation, just using them as a rationale to fund their domestic agri-business elites as domestic agricultural subsidies are cut (Woertz 2012)—then there is a strong motivation for them to intervene in ways that seek to enhance the likelihood of achieving those desired outcomes. For if GS investors—both the governments and their corporate proxies—are serious about achieving more reliable long-term sources of staple foods from investments in foreign agriculture, then investing in land in countries like Cambodia and Ethiopia is a flawed strategy. Instead they could be exploring an alternative: *investing in the farmers, not the land*. A measure of the true validity of the claims made that these investments are motivated by long-term food availability concerns and are truly seeking "win-win" outcomes will be whether investing country governments

[2] Discussion with Saudi Arabian policy maker in Doha, 12 November 2012. Also Alshareef (2009).

heed this message and create policy frameworks that encourage investment in least-developed country (LDC) farming communities instead of facilitating today's business-as-usual land deals.

Part of the power of the investing state governments is their ability to regulate and modify the behaviours of enterprises domiciled in their country. For companies proactively supported by the government, the regimes have direct financial leverage over those organizations and the ability to limit or halt support if expected codes of behaviour are not adhered to. Investor state governments also bargain with host countries and enter such negotiations from a position of strength. By tying investments to certain objectives—for example granting secure land tenure to peasant farmers they might wish to contract with—they have the capacity to affect change in the host country environment as well. As wealthy and powerful nations, engaged in multilateralism in a variety of forms, they are also in strong positions to set good examples and influence incremental normative shift at the international level. Thus, the policy suggestions put forward focus primarily on the investor state governments rather than the host state regimes or international organizations.

8.3.1 Vietnam: Success from Focusing on Smallholder Agriculture

Although it is contrary to dominant discourses that promote the industrialization of agriculture and agglomeration of farmlands, making investing in and promoting smallholder agriculture is a strategy that has been highly successful in Vietnam. Vietnam's rice industry turned from a situation of net food importation in the late 1980s to a net exporter by the early 1990s (Pingali et al. 1997). At the same time, the country went from being largely food insecure in terms of high levels of populations of vulnerable to hunger, to largely food secure with the vast majority of the population having access to both subsistence food and, albeit often small, income for additional needs (Jaffee et al. 2011). Since the 1990s Vietnam has grown to become the world's second largest rice exporter. The Vietnamese case is particularly interesting because of the change in government policies in the late 1980s that lead to its success. Consensus is that the single most important policy change was the granting of secure land-use rights to smallholder farmers.[3] Other crucial factors included opening the farmers' access to markets, large-scale investment in irrigation infrastructure, the provision of inputs and dissemination of know-how. The successes of this strategy were twofold. First, it resulted in the production of considerable surpluses by the smallholder farmers that were available for wider domestic consumption and for export. The latter generated substantial and reliable foreign exchange returns for the government, being largely government-controlled trade. Second, the strategy ensured at least a modest income for the vast majority of

[3] Interviews with policy -makers, -advisors and -analysts, Hanoi, 3–15 February 2012.

agrarian families. This saw dramatic improvement in Vietnam's overall food security situation since most families had adequate land for subsistence and at least a modest income. To illustrate: Vietnam reduced poverty between 1985 and 1995 from 40 % of the population to 20 % and by 2000 it was down to around 11 % (Le and Nguyen 2002).

The Vietnamese government granted land-use rights to smallholder farmers in the form of 20-year lease terms, although the taxes or fees in return for the leases are negligible. There was a concerted effort to distribute the land use rights fairly, so both "good" and "bad" paddy lands were divided amongst local families so that all got equivalent areas of each, although their plots are frequently non-contiguous. This has resulted in considerable fragmentation of land but relative equity amongst the farmers in any particular area. However plot sizes are small; especially in the Red River Delta where total allocations are often less than half a hectare per family (Jaffee et al. 2011). As the 20 year lease terms come up for renewal, the government is considering extending the lease terms to perhaps 50 years or longer.[4] Combined with the certainty of being able to sell their production—the government committed to buying as much rice was produced—the security of land tenure gave rice farmers the incentive to invest in the land and maximise their production. This certainty in the rice sector flowed on to create equivalent benefits in other farming sectors by ensuring domestic markets for other subsistence agricultural produce. The government also supported other export and cash crops (such as coffee and sugarcane) as it did rice (Hoa and Grote 2004). A technical expert described the importance of the land tenure reform to the success of Vietnam's agricultural sector with the following comparison:

> It is interesting to compare Vietnam and the Philippines. The Philippines got IRRI [the International Rice Research Institute]—now 50 years old—which was intended to solve food insecurity in Philippines, Vietnam and Indonesia through technical innovation. Vietnam, without IRRI, has been successful. The Philippines, despite IRRI, has not. This suggests that land rights for smallholders are more important than research and technology, in some cases. Indeed, you could say that research and technology are only effective once that land rights have been secured. Incentive first, then technology.[5]

8.3.1.1 The Vietnam Foreign Investment Regime

Contrary to the foreign direct investment (FDI) model the GSs are engaging in, the successes of the Vietnamese agricultural sector have been achieved mainly with domestic, public investment. Criticized by foreign capital as being "too difficult" and "not investment friendly,"[6] only 4 % of FDI in Vietnam is in agriculture. Vietnamese foreign investment rules require firms:

---

[4] Interview with Policy Advisor, Hanoi, 8 February 2012.

[5] Interview with Director of an agricultural research institution, Hanoi, 7 February 2012.

[6] Interview with Director of an agricultural research institution, Hanoi, 7 February 2012.

- With any foreign capital exceeding 30 % to operate under license from the Vietnamese government (which has a limited duration)
- To employ only Vietnamese staff and management and train Vietnamese to replace any foreign staff used in setting up the enterprise
- To buy Vietnamese inputs in preference to foreign inputs "where technical and commercial conditions are similar"
- To prioritise Vietnamese investors in future equity transactions pertaining to the enterprise
- To use Vietnamese insurance companies and auditors
- To bank with Vietnamese banks in Vietnamese currency; and,
- To transfer all technology brought into the enterprise from outside to Vietnam. (Socialist Republic of Vietnam 1996)

Although this approach is criticized by the transnational agribusiness corporations—including reportedly making the state the target of an aggressive campaign by the transnational agribusinesses at Davos in 2011[7]—it has not curtailed highly successful foreign investment and development in other sectors of the Vietnamese economy (Leproux and Brook 2004).

The importance of the social impacts of development and the prioritisation of agrarian peasants over corporate and other interests has been part of Vietnam's reluctance to allow the large transnational agribusiness corporations into the country. Recently, the government has partially relented to foreign pressure and granted concessions (although exactly what these entailed was not explained) in five sectors to five transnational corporations[8]:

- Tea – Unilever
- Coffee – Nestlé
- Vegetables – Pepsi-Cola
- Aquaculture – Cash & Carry
- Maize and soy – Monsanto

The policy maker explaining the background to these concessions to me expressed grave concerns, especially about Monsanto, who are now allegedly putting pressure on the Vietnamese government to retract its long-standing policy against genetically-modified organisms.[9] The possibility of equitable and successful rural-led development without transnational corporations as demonstrated by Vietnam thus far provides a constructive example for investment in smallholder agriculture instead of large-scale land deals.

[7] Interview with Policy Advisor, Hanoi, 8 February 2012.

[8] Interview with Policy Advisor, Hanoi, 8 February 2012.

[9] Interview with Policy Advisor, Hanoi, 8 February 2012.

8.3.1.2 Limitations

However, the Vietnam model is not perfect. There remain families and groups excluded from the benefits, especially indigenous communities in the uplands not engaged in conventional farming.[10] There also remain in Vietnam a few large state-owned farms from the period of soviet-structured agriculture. These were described being as "usually completely inefficient."[11] More importantly, over time, limitations to the model have become evident; especially as population continues to swell, the area of land per person in rural areas shrinks, and the small plot sizes limit the ability of farmers to increase their incomes as the rest of the economy grows (Jaffee et al. 2011).[12] Increasing industrialization encroaching on paddy lands is also negatively impacting land availability and plot size.

One of the challenges facing Vietnamese policy-makers today is how to facilitate diversification of the rural economy to promote non-agricultural employment and allow the more efficient smallholders to gradually increase their farm sizes while others reduce their direct dependence on the land for their livelihoods.[13] The Vietnamese policy officials I met with however do not see that a rapid transition to large-scale agglomeration of farmlands or corporatization of farming as a suitable strategy to achieve this. As demonstrated by the foreign investment regime described above, a preoccupation with Vietnamese officials seems to be concern with achieving largely equitable development, not development at the expense of Vietnam's large agrarian class.

Another (probably unexpected) outcome of the Vietnamese strategy has been quantity-at-the-expense-of-quality in Vietnam's rice production. This is in some part due to the difficulties of assuring high quality control across a vast number of producers. Some producers simply don't have the same capacity as others; for example, know-how, inputs, paddy land quality. But also, the quality of an entire truckload is only as good as the lowest quality in that batch meaning that conscientious farmers are penalized or discouraged from investing in higher quality practices by the poor performance of the negligent. This has meant that the majority of Vietnam's rice exports are at the low end of global market prices, further limiting the ability for the farmers (and also the state) to gain better returns. This quality issue, and a problem of ongoing post-harvest losses—the latter is not unique to Vietnam—are key concerns of policy makers today.[14] One aspect of the problem is infrastructure. While Vietnam invested heavily in irrigation that undoubtedly has been part of its success, other infrastructure such as local grain storage for farmers

---

[10] Interviews with NGO, Hanoi, 8 February and International Donor Institution, Hanoi, 9 February 2012.

[11] Interview with Academic and Policy Advisor, Hanoi, 10 February 2012.

[12] Also, Interviews with International Donor Institution, Hanoi 9 February 2012 and Policy Advisor, Hanoi, 8 February 2012.

[13] Interview with Director of policy think-tank, Hanoi, 7 February 2012.

[14] Interview with Policy Advisor, Hanoi, 8 February 2012.

remains under-developed. The impact of this lack is that farmers have to sell their crop immediately; usually at time of market glut and low prices. This limits farmer's flexibility and, ultimately, their incomes.

A consequence of the overwhelming focus on rice and the quantity-over-quality problem has been a tendency for the system to focus on production targets and export volumes rather than diversity and value. This creates some long-term risk—all eggs in one basket—and it was suggested that this has lead to an inclination to address weaknesses in the system by government intervention (price setting, buying up excess production, handouts etc.) which is not sustainable over the long term.[15] However this also creates opportunities: under-exploited market segments and possibilities for consolidation and vertical and horizontal integration for small to medium enterprises in the agricultural sector.

8.3.1.3 Lessons

Although Vietnam has not achieved its success via foreign commercial investment in its agricultural sector, there are lessons to be learned from Vietnam both from its successes and the limitations that have been encountered. It is possible to apply these lessons from the Vietnamese experience to developing suitable policies and investment practices that could encourage similar positive outcomes in Cambodia and Ethiopia from foreign, commercial investments in their agricultural sectors.

The opportunities for countries like Cambodia and Ethiopia to generate substantial improvements in agricultural productivity are evidenced by low production levels and the dearth of productivity-improving resources (Shepherd 2012a, b). The latter paucities range from irrigation to mechanisation and from inputs such as fertilizer to the knowledge of how to use them most effectively. The Vietnamese experience supports the case that the Cambodian and Ethiopian agricultural sectors present the possibility of generating surpluses for export via smallholder farmers. In comparison to Cambodia, Vietnam is very densely populated, and while Vietnam has the lion's share of the ultra-fertile Mekong River Delta, Cambodia's Mekong and Tonle Sap riparian zones have plenty of underutilized agricultural potential, producing considerably less than neighboring regions on the other side of the border in Vietnam.[16] This further supports the notion that there is hope for productivity increases if Cambodian farmers can get support similar to that received by Vietnamese farmers.

---

[15] Interview with International Donor Institution, Hanoi, 9 February 2012.

[16] According to Cambodian government figures, only 16 %—385,000 ha—of the 2.4 million hectares of paddy land are cropped during the dry season while just 4 % of those 2.4 million ha produce two crops over the wet season: (Supreme National Economic Council [SNEC] 2010) This compares with three crops per year being grown by smallholders in An Giang, one of the most productive of Vietnam's rice growing provinces, which lies across the Cambodian border and along the (shared) Mekong River: (Doan et al. 2010).

The Vietnam situation also affords hope for Ethiopia. Ethiopia's famously bad food security situation and its large, poor, population suggests that achieving domestic food security will be difficult. While Vietnam is blessed with its fertile rice lands, it has a population approximately the same as Ethiopia's and actually considerably less total arable land (FAOSTAT 2011). This suggests that on a person-per-arable-hectare basis the Ethiopian situation is not as hopeless as the its grim history of food insecurity might imply. Furthermore, Vietnam like Ethiopia has a land system where the state legally owns all the land. In Ethiopia, this has been a barrier to provision of secure land rights to farmers and has been argued to be a significant factor in the weakness of Ethiopia's agricultural sector (Rahmato 2004).[17] The land-use policy reform in Vietnam however demonstrates how provision of farmer tenure security does not need to be in contradiction with the fundamental position of the government that all land is (technically) owned by the state. The Ethiopian government could in theory, replicate the land-use rights granted to Vietnamese farmers without relinquishing that policy which it sees as an uncrossable "line in the sand."[18] Ethiopia also has an immediate advantage over both Cambodia and Vietnam in terms of avoiding the problem of monoculture and excessive focus on a single commodity. Ethiopia's relatively diverse agriculture offers opportunities for investing in smallholder production of multiple staple crops.

Both Cambodia and Ethiopia could benefit from immediate policy focus on smallholder farmers to kick-start development—and better food security—as Vietnam did in the late 1980s. If either government was serious about pursuing the opportunities in domestic agriculture, it could follow Vietnam's original strategy immediately and develop mitigations for the limitations of that model over time. It could also seek to develop mitigations from the outset, for example by promoting local small-scale processing and trading initiatives at the same time as securing farmers' land-use rights which would, as well as creating access to markets, contribute to bolstering off-farm employment in the rural areas. The Cambodian and Ethiopian governments could also learn from the Vietnamese approach to foreign investment regulation and its cautious treatment of transnational agribusiness enterprises in the country. However, as has been assayed earlier, it is unlikely that the Cambodian or Ethiopian regimes are likely to adopt such policy suggestions and, instead, the opportunity to affect change lies with the investor states.

8.4 For Gulf State Policy-Makers

The first, and probably most obvious, lesson for Gulf State policy makers is that security of land tenure is crucial for farmers. Indeed, the preoccupation of investors in securing long-term leases for their own confidence to invest in improving the

[17] Also discussed Shepherd (2012b).

[18] Interview with Policy Maker, Addis Ababa, 14 October, 2012.

productivity of the land suggests that this should not come as a surprise. What is also important is that land tenure security is accompanied by access to markets for the farmers' production. As noted above, technology and infrastructure are important, but secondary to giving farmers certainty and incentive: Vietnam has been successful even with some important infrastructure still lacking.

Investor states by their very participation can provide access to markets for LDC agriculture. The provision of know-how, infrastructure and inputs (or at least in the context of smallholder farming, the ability for farmers to access cheap credit for purchasing inputs) are what an investor would be bringing to the LDC agriculture environment in any circumstance (under any type of investment model). A key question then is what a foreign commercial investor is able to offer in terms of land tenure security. Unlike those other things, providing certainty to farmers over access to land, however, is contrary to the expectation of investors who have thus far been seeking long-term land tenure for themselves. But as argued elsewhere, land taken by investors at the expense of local communities ultimately jeopardises the viability of agricultural projects. Instead, investors helping to secure local communities land-use rights may well be a solution to mitigating those risks.

There are two ways this could work in practice. The first is that when investor state governments like the Saudi Arabian regime are discussing in-country investments with the host country government, they negotiate deals on the basis of long-term land tenure security being granted to the communities in the areas being considered for projects. The second is that investors sign long-term land leases with the host country regime that the investor subsequently contracts back to the local communities in the project area for the entire duration of the investors' lease contract with the government. The purpose of this is to pass the secure usage rights the government has granted the investor over to the famers instead.

Both of these have weaknesses. The primary weakness of the former is the risk that secure land tenure is awarded by the host government to politically acceptable portions of the local community at the expense of politically disposable elements, still leaving significant portions of the community disgruntled and excluded. Given the situation in Ethiopia,[19] this is a foreseeable possibility there, although it would seem to be less likely in Cambodia where land is not employed for political purposes in the same ways it is in Ethiopia. It would be feasible to negotiate specific terms within the government-to-government contracts with monitoring mechanisms to curb this possibility. In reality, there are limits to which negotiators are willing to be seen to interfere in the domestic affairs of the country with whom they are negotiating which might make such mechanisms unlikely. The primary weakness of the latter is the difficulty of executing such back-to-back agreements in the absence of farmer co-operatives or other rural social organizations with which the investors can negotiate. This is one area where foreign investors probably cannot create change directly, however the very presence of prospective investors—and demonstrating the goodwill to engage with such organizations—may stimulate the motivation for their

[19] As described in Shepherd (2012b).

establishment. It must also be reiterated that while the "incentive" aspects have been prioritized here, this does not diminish the "technical" aspects that investors also can, and must, bring to the table as well.

8.4.1 Recommendations: Rethinking Food Security Policies

To draw some recommendations for investor states' policy makers from this discussion, it is worth extracting two principles from the experience of the land deals as they stand and the opportunities and benefits conferred by rethinking them in the light of food security as securing the vulnerable from hunger. These principles provide the rationale for the recommendations; if the investor buys in to the principle, then there is no reason why the recommendation could not be adopted. The first principle is that poor investment practices create risks for long term food security goals if those practices dispossess, injure or impoverish local communities. That poor investment practices easily result in dispossession, injury and impoverishment of local communities has been comprehensively exhibited by the research in Cambodia and Ethiopia. The second principle, then, is simply that agricultural development investments that empower and create value for local communities are likely to be more successful over the long term than those which confiscate resources from those communities.

The first recommendation for GS (and indeed other investor state) governments to draw from these principles is that—prior to embarking on any further projects—the states need to (either individually or collectively) regulate the extraterritorial activities of their businesses. This suggestion is not without precedent. John Ruggie is the Special Representative of the United Nations Secretary-General on the issue of human rights and transnational corporations and other business enterprises. The Ruggie Report, *Guiding Principles on Business and Human Rights: Implementing the United Nations "Protect, Respect and Remedy" Framework,* seeks to create a normative benchmark for responsible behaviour of transnational capital. Ruggie's position is that states must "set out clearly the expectation that all business enterprises domiciled in their territory and/or jurisdiction respect human rights throughout their operations" (Ruggie 2011, 7 Foundational Principle No. 2). There are no sound reasons why the GSs and others with enterprises investing in LDC agriculture could not follow this prescription and implement the Ruggie framework in its entirety as a strategy to help encourage good investor and enterprise behaviour. In the mean time, a moratorium on land-acquiring investments could be instituted until mechanisms are in place to mitigate the risks and pathologies arising from them.

The second recommendation is for GS regimes to change their negotiation strategies when dealing with host countries over investment programs. The investors are in a strong negotiating position. Contrary to the views currently held by many GS policy makers, their interests are best served not about securing land-use rights but about securing investment partnerships that give them long-term reliable access to staple food production. They could use their negotiating strength to institute blanket terms with host governments that include land rights for local farmers. They could also help codify expectations of behaviour and performance of their own firms in that country, where such behaviour is geared towards sustainable production of surpluses for export by the smallholders.

To this end, in addition to—and more specifically than—signing up to the Ruggie framework, the investor state governments could develop a clear regulatory model for their investing enterprises to adhere to, to maximise the likelihood of successful ventures in LDC agriculture, in line with the blanket agreements negotiated with host country governments. This is the third and most important recommendation. Although a legal model can be linked to government support of investing enterprises as a condition of the support, it can also be used to regulate independent investment ventures to ensure that "rogue" corporate activity does not damage the efforts to develop a successful investment regime by association. Important aspects of a regulatory model are that:

1. It adopts a framing of food security which prioritizes securing the vulnerable from hunger.
2. Investment projects engage with local communities to bolster their land-use rights, and do not obtain land-use rights for the investor.
3. Investors ensure market access at fair prices for farmers' surplus production. Fair prices can be established by mechanisms such as a percentage of FOB prices on a regional mercantile exchange (for example 80 % of the average Bangkok FOB rice price for the week prior to the date of transaction).
4. Investors commit to providing key infrastructure required by the communities to lift their productivity.
5. Investors make low-cost finance available for inputs and facilitate dissemination of know-how and access to low-cost inputs (for example with the assistance of import advantages offered to foreign companies in both Cambodia and Ethiopia not otherwise available to local farmers).
6. It prioritizes infrastructure developments and investment in the provision of inputs and know-how that are ecologically and socially sustainable in the LDC.
7. The government establishes independently audited public accountability mechanisms for all projects, including pricing mechanism reviews, environmental impact monitoring and farmer grievance handling.

The first point does not negate that investor enterprises and the investor states are seeking reliable (and profitable) supplies of staples: this is a business imperative that will be pursued by business interests anyway. What it is ensures is that in the pursuit of those objectives, priority is given to the needs of the poor and vulnerable. Points number 2 and 3 contribute to creating the incentive for local smallholder farmers in the LDC. Items 4 and 5 aim to provide the technical capacity to facilitate the smallholders maximising their productivity. Point 6 is crucial to avoid investors seeking short-term returns that mine soils and deplete fresh water resources which will thwart the objective of the host country providing reliable surplus productivity for the long term. The final point, 7, might seem counter-productive to usual business and investment practices that tends to shirk transparency. However, it is, in actuality, crucially important for monitoring the types of risks that jeopardize the long term viability of these projects: dispossession, injury or impoverishment in the host country. It is these which must be prudently avoided if food security projects in foreign countries are to have high prospects for long-term success. Perhaps more importantly, the provision of such mechanisms ensures that the most important stakeholders—the farmers who will be generating the long-term supply of staple produce—are kept engaged, adequately rewarded and, consequently, productive.

8.4.2 Investment Partnerships, Not Contract Farming

A salient feature of all the above points is that they involve the farmers—ideally, though not essentially, represented by cooperatives or other form of social organisation—in the host country. It would be easy but incorrect to equate the model discussed above with contract farming. Contract farming in least developed countries is usually largely unregulated (or if regulated, those regulations are largely unenforced) and can tend to be in the form of deeply unequal contracts between uneducated and disempowered farmers and rich and savvy corporate enterprises (Glover and Kusterer 1990; Watts and Little 1994).

To illustrate these criticisms, a telling example of the weaknesses in contract farming in developing country environments came to light while I was in the Philippines during the course of this research. In this particular situation a corporate investor from an Arab state on the Arabian Gulf, operating under the auspices of its government's external food security policy, made a 25-year deal with a local businessman in the Philippines to secure the production from a co-operative of farmers for the purposes of exporting their produce back to the Gulf and other profitable markets including Japan. Viewed from the conventional perspective of food security as a food availability problem, this case appears laudable on a number of fronts. First, it superficially supports its own government's initiatives in building "food security" for the Gulf state by going abroad to secure food-producing resources. This is despite the small matters that bananas are not really an essential staple foodstuff for the domestic populations of the Gulf States and that sales to Japan purely for profit are not increasing the amount of food available in the company's home market. Second, it supports the Filipino government's strategy to build domestic "food security" by increasing investment in its agricultural sector, based on the neoliberal premise that economic growth is a rising tide that lifts all boats. Third, the deal has been made with a farmers' co-operative in Mindanao who are successful beneficiaries of the Filipino land reform program and have title to their own farmlands. Within conventional framing, this demonstrates the success of the state's land reform programs, enabling farmers freedom over the employment of their land-capital, increasing their economic (and hence food) security (Flores-Obanil and Manahah 2007; Shepherd 2011).

However, when approaching this same case with the perspective of a re-framed understanding of food security as protection of the vulnerable from the risk of hunger and poverty, a different set of questions need to be asked and a different understanding of the situation results. On the assumption that the farmers were not actually suffering hunger at the time the deal was made, a central question is *how has this arrangement protected the rural farming communities from the lack of capacity in the face of risk to their livelihoods?* That is, *what is their protection from the threat of future hunger?* And, *how has the deal with the businessman and foreign investor increased their resilience to hunger?* In regards to the latter, the long answer is that the long-term lease-back arrangement pays the farmers between PhP15,000–20,000 (US$350–450) per hectare per year and in return for this annual income the farmers work the land on behalf of the firm who owns—and takes—100 % of the lands' production. Out of that lease-income the farmers are responsible for all input costs (pesticides, fertilizers and debt repayments to the state Land Bank for the reformed land) and have volume and quality obligations to meet. While this arrangement has granted a long-term, steady income for the farmers (but unchanging; so actually declining over the long-term in the context of a growing Philippine economy), it has effectively excluded them from the full benefits of their land-capital because they have effectively surrendered the ability to use the land, its produce or its financial value, for any other purpose. This potentially may have included sale of the produce from the standing crop on the open market; subsistence cropping, alternative—potentially more profitable—crops; mortgaging the land for funds to invest elsewhere; or sub-leasing the land to other producers. Thus one of the roles of the foreign investor and local businessman in this situation of contract farming appears to have been to *reduce* the capacity of the farmers by committing them to a long-term, low-income future and removing the options of pursuing alternatives. Digging further into this case, the Middle Eastern investor claimed to be making a claimed 300 % higher profit margin from this particular venture than their previous venture in the same locality which was a trading-only business that bought the banana crop at market prices from farmer co-operatives to transport and sell into its export markets: in terms of a protective understanding of food security the former situation would appear to be a more successful programme than the latter.[20]

In some respects the contract farming example above is not so different from land deals which exclude the farmer from the land, instead they are effectively indentured to the investor for a lifetime of work with little return. Perhaps the differences between the investment partnerships being proposed here and contract farming as illustrated by the above example are subtle. However, they are important. The key difference is the empowerment of the rural poor; not only by shifting the framing of food security to focus on addressing their vulnerabilities to hunger, but recognising the role that they can—and must play—in delivering the objectives desired by the investors for the long haul. More than contracting with farmers,

[20] Interview with the local agent of the GS investor firm, Davao, The Philippines, 15 October 2010.

investment partnerships are not about contracting over the productivity of the land—where the farmers end up carrying the risk—but about investing in farming communities to develop the productivity of the land for the mutual benefit of both parties well into the future.

8.5 Conclusion

Previous research has identified some significant pathologies with acquisition of usage rights over agricultural land in least developed countries as a strategy by GSs and others to secure long term food supplies. These pathologies include dispossession and impoverishment of local communities—creating food insecurity instead of food security—and the risk of fuelling of conflict. These in turn make the achievement of food security goals from such investments unlikely. This chapter has examined the proposal that investment partnerships between GS investors and smallholder farming communities in LDCs could provide better food security outcomes for both the investor and the LDCs. It has used the example of Vietnam with its successful strategy of focusing on smallholder agriculture to generate surpluses for export as well as limiting food insecurity in rural areas. In light of the Vietnamese experience, this chapter has put forward some recommendations for GS policy makers to limit the pathologies of existing land-centric investment practices and to encourage the investment partnership model instead. It is suggested that if GS policy makers are truly seeking reliable sources of staple foods for the long term, then they could thoroughly consider the impacts of current strategies and invest in further research to demonstrate, in practice, the validity of the ideas put forward in this chapter as a more productive strategy than the current one of land deal investments.

References

Alshareef T (2009) King Abdullah's initiative for agricultural investment abroad. Kingdom of Saudi Arabia Ministry of Commerce & Industry. http://www.agritrade.org/events/documents/Alshareef.pdf

Anseeuw W, Boche M, Breu T, Giger M, Lay J, Messerli P, Nolte K (2012) Transnational land deals for agriculture in the Global South. http://landportal.info/landmatrix/media/img/analytical-report.pdf

Ayoob M (1995) The third world security predicament: state making, regional conflict and the international system. Lynne Rienner, Boulder

Bladd J (2010) ME's farmland buys in Africa seen as 'a win-win partnership'. Arabian Business, 15 April. http://www.arabianbusiness.com/586105-gulfs-farmland-policy-seen-as-win-win-partnership-for-africa

Blas J (2008) Foreign fields: rich states look beyond their borders for fertile soil. Financial Times, 19 August. http://www.ft.com/cms/s/0/8de8a3e0-6e17-11dd-b5df-0000779fd18c,dwp_uuid=5cdb1d20-feea-11db-aff2-000b5df10621.html

Borras SM, Hall R, Scoones I, White B, Wolford W (2011) Towards a better understanding of global land grabbing: an editorial introduction. J Peasant Stud 38(2):209–216. doi:10.1080/03066150.2011.559005

Bueno des Mesquita B, Smith A, Siverson R, Morrow J (2005) The logic of political survival. MIT Press, Cambridge

Cotula L, Vermeulen S, Leonard R, Keeley J (2009) Land grab or development opportunity? Agricultural investment and agricultural land deals in Africa. FAO, IIED & IFAD, Rome

Daniel S, Mittal A (2009) The great land grab: rush for the world's farmland threatens food security for the poor. http://www.oaklandinstitute.org/sites/oaklandinstitute.org/files/Land Grab_final_web.pdf

Doan NP, Nguyen HA, Nguyen HH, Singleton G, Palis F, Johnson D, Lampayan R, Gummert M, Duong VC (2010) Applying good agricultural practices to reduce production costs and environmental footprint in the Mekong Delta—the An Giang model. Paper presented at the 28th international rice research conference, OP02: closing the yield gap, 8–12 Nov, Hanoi, Vietnam. http://ricecongress.com/previous/extPdfs/OP02-3773-Pha-edited.pdf

England A, Blas J (2009) Qatar SWF to target food and energy. Financial Times. Available online: http://farmlandgrab.org/2827. Accessed 18 May 2010

FAOSTAT (2011) Statistical database of the food and agriculture organisation of the United Nations. http://faostat.fao.org/default.aspx

Federal Democratic Republic of Ethiopia [FDRE] (2010) Population and housing census report-country – 2007. Central Statistics Agency, Addis Ababa

Flores-Obanil C, Manahah MA (2007) Leaseback arrangements: reversing agrarian reform gains in the Philippines. http://www.landaction.org/spip/spip.php?article17

Global Witness (2007) Cambodia's family trees: illegal logging and the stripping of public assets by Cambodia's elite. http://www.globalwitness.org/sites/default/files/pdfs/cambodias_family_trees_low_res.pdf

Glover D, Kusterer K (1990) Small farmers, big business: contract farming and rural development. St. Martin's Press, New York

Grain (2009) Rice land grabs undermine food sovereignty in Africa against the grain, Jan 2009. Available online: http://www.grain.org/articles/?id=46. Accessed 15 May 2010

Gustavsson J, Cederberg C, Sonesson U, van Otterdijk R, Meybeck A (2011) Global good losses and food waste. Food and Agriculture Organisation of the United Nations, Dusseldorf

Hoa N, Grote U (2004) Agricultural policies in Vietnam: producer support estimates, 1986–2002, ZEF discussion papers on development policy no. 93

Jaffee S, Sanh NV, Dung LC, Loc VTT, Nhan DK (2011) From rice bowl to rural development: challenges and opportunities facing Vietnam's Mekong Delta region. World Bank, Hanoi

Kingdom of Saudi Arabia Ministry of Foreign Affairs (2010) Custodian of the two holy mosques receives Minister of Commerce and Industry, Riyadh. Originally posted at Durable link at http://www.farmlandgrab.org/post/view/2717

Le TH, Nguyen CK (2002) Malnutrition and poverty alleviation in Vietnam during the last period 1985–2000. Asia Pac J Clin Nutr 11(Suppl):S331–S334

Leproux V, Brook D (2004) Vietnam: foreign direct investment and post crisis regional integration. Asian Development Bank ERD working paper no. 56. http://www.adb.org/sites/default/files/pub/2004/wp056.pdf

Monbiot G (2008) Snatching food from the mouths of the poor. The Guardian, 26 August. http://climateandcapitalism.com/?p=516

National Committee for Population and Development (2010) Cambodian population datasheet. Office of the Council of Ministers, Royal Government of Cambodia, Phnom Penh

Pingali P, Khiem NT, Gerpacio R, Xuan VT (1997) Prospects for sustaining Vietnam's reacquired rice exporter status. Food Policy 22(4):345–358

Rahmato D (2004) Searching for tenure security? The land system and new policy initiatives in Ethiopia. Forum for Social Studies, Addis Ababa

Rahmato D (2011) Land to investors: large-scale land transfers in Ethiopia, FSS policy debates series. Forum for Social Studies, Addis Ababa

Reuters (2008) Kuwait eyes Asia for food supply, agriculture. Reuters. Available online: http://farmlandgrab.org/2457. Accessed 18 May 2010

Reuters (2009) Kuwait eyes African investments. Reuters. http://www.tradearabia.com/news/_164576_201917.html or durable link at http://www.farmlandgrab.org/post/view/6339-kuwait-eyes-african-investments
Ruggie J (2011) Guiding principles on business and human rights: implementing the United Nations "Protect, Respect and Remedy" framework. New York. http://www.business-humanrights.org/media/documents/ruggie/ruggie-guiding-principles-21-mar-2011.pdf
Sachs J (2006) The end of poverty. Penguin, New York
Sen A (1981) Poverty and famines: an essay on entitlement and deprivation. Oxford University Press, Oxford
Sen A (ed) (1990) The political economy of hunger. Clarendon, Oxford
Sen A, Dreze J (1987) Hunger and public action. Clarendon, Oxford
Shepherd B (2010) Above carrying capacity: Saudi's external policies for securing food supplies. Paper read at World Congress of Middle Eastern Studies, Barcelona, 19–24 July 2010. http://www.benshepherd.com.au/wp-content/uploads/2014/06/Shepherd-Above-Carrying-Capacity-WOCMES-2010.pdf
Shepherd B (2011) Redefining food security in the face of foreign land investors: the Philippine case. Centre for Non-Traditional Security Studies. http://www.rsis.edu.sg/nts/HTML-Newsletter/Report/pdf/NTS-Asia_Ben_Shepherd.pdf
Shepherd B (2012a) GCC States' land investments abroad: the case of Cambodia. Summary report no. 5. http://www12.georgetown.edu/sfs/qatar/cirs/CambodiaSummaryReport.pdf
Shepherd B (2012b) GCC States' land investments abroad: the case of Ethiopia. Summary report Georgetown University Centre for International and Regional Studies. http://www12.georgetown.edu/sfs/qatar/cirs/TheCaseofEthiopiaSummaryReport.pdf
Shepherd B (2012c) A need for new institutions? Bilateral deals over food producing resources point to the need for greater international cooperation on food security. In: Krishna-Hensel SF (ed) New security frontiers. Ashgate, Farnham
Shepherd B (2012d) Thinking critically about food security. Secur Dialogue 43(3)
Socialist Republic of Vietnam (1996) Law on foreign investment in Vietnam. National Assembly, Hanoi. http://www.vietnamlaws.com/freelaws/LFIna12Nov96%28aa9Jun00%29%5BI1%5D.pdf
Speildoch A, Murphy S (2009) Agricultural land acquisitions: implications for food security and poverty alleviation. In: Kugelman M, Levenstein S (eds) Land grab? The race for the world's farmland. Woodrow Wilson International Centre for Scholars, Washington, DC
Supreme National Economic Council [SNEC] (2010) Policy paper on the promotion of paddy production and rice export. Council of Ministers, Royal Government of Cambodia, Phnom Penh
Tetreault MA, Wheeler D, Shepherd B (2012) Win-win versus lose-lose: investments in foreign agriculture as a food security strategy for the Arab Gulf states. In: Babar Z (ed) Food security and food sovereignty in the middle east. Stanford University Press, Palo Alto
Uvin P (1994) The international organisation of hunger. Kegan Paul International, London
Walid T (2009) Qatari SWF food unit eyes PAVA stake, seals Sudan deal. Reuters. Available online: http://af.reuters.com/article/investingNews/idAFJOE59S0LW20091029?pageNumber=1&virtualBrandChannel=0. Accessed 22 May 2010
Watts M, Little P (eds) (1994) Living under contract: contract farming and agrarian transition in Sub-Saharan Africa. University of Wisconsin Press, Madison
Woertz E (2011) Arab food, water and the big gulf landgrab that wasn't. Brown J World Aff 18 (Fall/Winter):119–132
Woertz E (2012) Oil for food: the global food crisis and the middle east. Oxford University Press, Oxford
World Bank (2010) Rising global interest in farmland: can it yield sustainable and equitable benefits? http://go.worldbank.org/XWESRO2MT0

Chapter 9
Impact of Food Prices, Income and Income Distribution on Food Security in Oman

Hemesiri Kotagama, Salwa Abdullah Nasser Al Jabri, Houcine Boughanmi, and Nejib Guizani

Abstract The impact of the surge in food price in 2008 and changes in income and its distribution on food security in the Sultanate of Oman is analyzed. The threshold of household food security was defined as access to a Nutritionally Adequate and Preferred Least Cost Diet (NAPLC). Changes in Food Security Head Count (F0: percentage of population unable to access NAPLC) and Food Security Gap (F1: a measure of amount of income that is required to bring all household that are food insecure to NAPLC) due to changes in food prices, income and its distribution were estimated using Software Platform for Automated Economic Analysis. With the surge in world food prices, Oman's consumer price index of all food increased by 21.60 % in year 2008 compared to 2003. The average household income increased from 638.00 to 913.00 Omani Rial (OR) per month per household and the income distribution has significantly improved towards equality with the Gini-coefficient changing from 46.49 to 36.35 from year 2000 to 2008, respectively. F0 was 24.0 % with food prices, income and its distribution as prevailed in 2003. Due to the increased food prices in year 2008, even with increased income and more equal income distribution F0 had increased to 29.3 %, increasing food insecurity by 5.3 %. Had the food prices not increased the increased income and changed income distribution towards equality, F0 would have improved to 9.7 % by 2008 which is an

---

H. Kotagama (✉) • H. Boughanmi
Department of Natural Resource Economics, College of Agricultural and Marine Sciences, Sultan Qaboos University (SQU), Al-Khod 123, PO Box 34, Muscat, Oman
e-mail: hemkot@squ.edu.om; boughnah@squ.edu.om

S.A.N. Al Jabri
Planning and Investment Promotion, Ministry of Agriculture and Fisheries Wealth, PO Box 467, Muscat PC 100, Oman
e-mail: salwa.aljabri@gmail.com

N. Guizani
Department of Food Science and Nutrition, College of Agricultural and Marine Sciences, Sultan Qaboos University (SQU), Al-Khod 123, PO Box 34, Muscat, Oman
e-mail: Guizani@squ.edu.om

S.A. Shahid and M. Ahmed (eds.), *Environmental Cost and Face of Agriculture in the Gulf Cooperation Council Countries: Fostering Agriculture in the Context of Climate Change*,
DOI 10.1007/978-3-319-05768-2_9,

improvement of food security by 14.3 % compared to F0 of 24.0 % in 2003. The 14.3 % improvement in food security would have been caused by almost equally through increased income (6.31 %) and improvements in the distribution of income (6.92 %) and due to interactive effect of increased income and improved distribution (1.09 %). F1 too was found to show similar changes in food security during 2003–2008 as F0. The average F1, amount of income required to bring all food insecure households to food secure threshold was estimated at 50.680 OR per month per household. The analysis indicates that food security in Oman would have significantly improved due to increased per capita income and improvements in the distribution of income had the food prices not surged in 2008. However the drastic increase in food prices in 2008 has instead increased food insecurity.

Keywords Food prices • Food security measures • Income distribution • Oman • Price surge

9.1 Introduction

Food security is a fundamental need for the improvement and sustenance of human welfare. Food security is defined as a situation when all people, at all times, have physical and economic access to sufficient, safe and nutritious food to meet their dietary needs and food preferences for an active and healthy lifestyle (FAO 2003). The world food security improved, since the green revolution in the 1940s–1970s. Since then the challenge of achieving food security has not been one of producing sufficient food for the growing population but a problem of distribution and access to food, which partly depends on income and its distribution (Leathers and Foster 2004). The surge and volatility of food prices since year 2008 has reawakened the need to examine food security, particularly in developing nations, where income is a dominant constraint to access food and in nations that are highly dependent on international markets for their food needs. Dawe and Morales-Opazo (2009) in analyzing Food and Agriculture Organization's data has reported a 48 % increase of world-wide food prices since 2008. Gilbert and Morgan (2010) in reviewing the most recent research literature on analysis of reasons for the food price surge, give prominence to rapid economic growth particularly in China and Asian economies, long-term under-investment in agriculture, low inventory levels, poor harvests due to droughts in major wheat producing countries, diversion of food crops to production of bio-fuels and market speculations as the causes of the world wide food price surge.

9.2 Oman's Food Security

The Sultanate of Oman is a sub-tropical country with limited water resources, harsh weather and low soil fertility, constraining food production. Therefore, Oman largely depends on international markets to assure supply of food. Between 2005 and 2007 Oman imported 44 % of the food consumed, 100 % of rice and about 95 % of wheat (MNE 2010a). Expenditure on food, is the largest percentage of the total household income which was about 31 % (MNE 2010b). In Oman a family is classified as poor if it spends more than 60 % of the household expenditure on food (MNE 2010b). Based on this standard 12 % of Omani families are classified as poor based on Household Expenditure and Income Survey conducted in 2007–2008 compared to 8 % in 1999–2000 (MNE 2010b).

9.2.1 Income, Its Distribution and Food Security

The basic causes of food insecurity in the modern world are poor purchasing power and lack of access to productive resources among the poor, high food prices and skewed income distribution (Senauer and Sur 2001). Timmer (2000) has empirically reasoned that achieving food security does not result from private decisions made in response to free-market signals alone and that food security could be achieved through government policies that seek economic growth with improved income distribution. Improvement in income distribution is critical in achieving food security of a nation since poor people may remain to be food insecure though national average consumption of food is adequate (Pinstrup-Anderson and Caicedo 1978; Gabbert and Weikard 2005). Pinstrup-Anderson and Caicedo (1978) have showed, based on an empirical analysis done in Colombia, that changes in income distribution can improve human nutrition, even in the absence of expansion of food supply.

9.3 Research Problem and Objectives

Food security depends on three factors: food availability, food accessibility and food utilization. In a mixed economy with market dominance, where most food is accessed from markets, such as Oman, household access to food depends on household income. Food prices through impacting real income of households tend to constrain access to food, leading to food insecurity. Oman has experienced substantial change in both growth in per capita income and its distribution. Understanding the impacts of changes in food prices, income and its distribution on food security, will help in designing future policies and strategies to improve and sustain food security in Oman. Thus the objective of this study is to empirically estimate the impact of the recent (2008) surge in food prices, the changes in income and its distribution on food security in Oman.

9.4 Literature Review

9.4.1 Global and Regional Food Security

Studies on food security were inspired by the seminal publication by Malthus (1798) prophesying the inevitability of food insecurity, given the geometric growth of population and arithmetic growth of food production. Although Malthus's prophesy did not become a reality, due to many factors, among which technological advance in food production is a prime factor, the challenge of achieving global food security remains. At present (2009) nearly one billion (1/6th of the world population) are food insecure. Hence there is consensus that world's food insecurity is not due to lack of resources or technology to produce sufficient food for the global population but due to failure in the equitable distribution of food among the world population. The dominant institution that distributes food is the market. Access to food in the market whether a nation or household depends on income and prices of food. An increase in food prices thus aggravates the constrained access to food among the poor. Food prices surged in the 1970s and most recently in 2008. Studies on the impact on increasing food prices (hence real income and purchasing power) on food security thus dates back to the 1970s and to the most recent studies post 2008 (World Bank 2009, 2010a, b; FAO 2008).

El-Sherbani and Sinha (1978) have observed that despite the decrease in the relative importance of the agricultural sector vis-à-vis the growth of the oil sector, it is still important for Arab countries to improve its agriculture due to possible cutoff of food exports by food exporting countries, for political reasons. The study thus highlights political factors as a unique but important cause of food insecurity. Adamowicz (1988) has identified the vulnerability of Arab countries to food insecurity, given the rapidly growing population (3 % in the 1970s–1980s) and disposable income (with income elasticity of demand for food between 0.2 and 1.0 for Arab countries) and the constraints of resources to produce particularly grains (agricultural production growth as 2.5 %). Food security has been considered as a problem of short term variability in national production and instability of imports. The possibility of a more egalitarian distribution of income to facilitate the achievement of food security has also been emphasized.

9.4.2 Food Security in Oman

Oman depends largely on imported food because of limited agricultural land, water resource and harsh climate. Further, many people in Oman shifted from employment in the agriculture sector to other employment after the discovery of oil. There has been a small change in the estimated area cultivated in Oman. The estimated cultivated area was 175,000 feddan (1 feddan = 0.42 ha) in 1999 and remained the same after 10 years with a small change in the estimated production (1,287,000 t in 1999 and 1,187,000 t in 2009) (MNE 2010a).

Table 9.1 Actual and recommended consumption of nutrients by an Omani household (per day per family)

Nutrient	Unit	Actual consumption/day	Recommended consumption/day	Percentage fulfillment
Energy	Kcal	18,893.3	15,644.6	121
Protein	g	600.2	312.9	192
Carbohydrate	g	2,944.2	2,148.5	137
Fat	g	551.2	860.5	64
Dietary fiber	g	176.3	125.0	141
Vit. B1	mg	6.8	7.8	86
Vit. B2	mg	8.6	9.3	92
Vit. B6	mg	12.7	7.8	162
Vit. C	mg	1,966.4	391.1	503
Calcium	mg	4,754.1	3,911.0	122
Iron	mg	80.7	86.1	94
Vit. B12	μg	19.9	7.8	255

Source: Al Jabri (2011)

As reported by Alasfoor et al. (2009) in the Food Based Dietary Guidelines for Oman, the average Omani household is nutritionally food secure with consumption of adequate amounts of energy and protein, whilst deficient in the intake of some micronutrients. A recent study (Al Jabri 2011) that compares the actual consumption with recommended consumption of food components of an average Omani family (eight members) confirms above (Table 9.1).

There are a number of nutrition interventions by Ministry of Health (MoH 2009) such as providing vitamin A supplements to the women and children, implementing food fortification programs like iodized salt and fortified white wheat with elemental iron and folic acid. The general health status in Oman is that among preschool children 9.5/1000 person are affected by Protein – Energy Malnutrition in 2008, about 8 % are underweight and 29 % are overweight. Among adults 17 % are obese (MoH 2009).

9.4.3 Income and Its Distribution and Food Security

Income and its distribution are major factors that determine household food security. Whilst increase in income is well understood to improve household food security the impacts of changes in income distribution on food security is not obvious. Senauer and Sur (2001) provide evidence in improvement of food security associated with improvement of income distribution. It is estimated that economic growth favoring the poor would bring down the world population that is food insecure to about 608 million (from 800 million in year 2000) by year 2025. As quoted by Senauer and Sur (2001) the World Bank (2000) considers that economic growth associated with investments on improving human capital, particularly education, health care and improving employment opportunities for the poor and in improving small holder agriculture improves food security through improved income distribution.

9.5 Analytical Methodology

The primary methodological requirement to achieve the objectives of the study is to measure households who are below a food security threshold due to changes in food prices, income, and its distribution. A food security threshold of Nutritionally Adequate and Preferred Least Cost Diet (NAPLC) for Oman's households estimated using a linear programming model by Al Jabri and Naser (2011) is used in this study. Distribution of Disposable Income for Food (curve DDIF in Fig. 9.1) is derived from income distribution data of Oman and by considering the proportion of income spent on food by different income levels. The households falling below the food security threshold (NAPLC) is measured considering the DDIF distribution curve (area FI in Fig. 9.1).

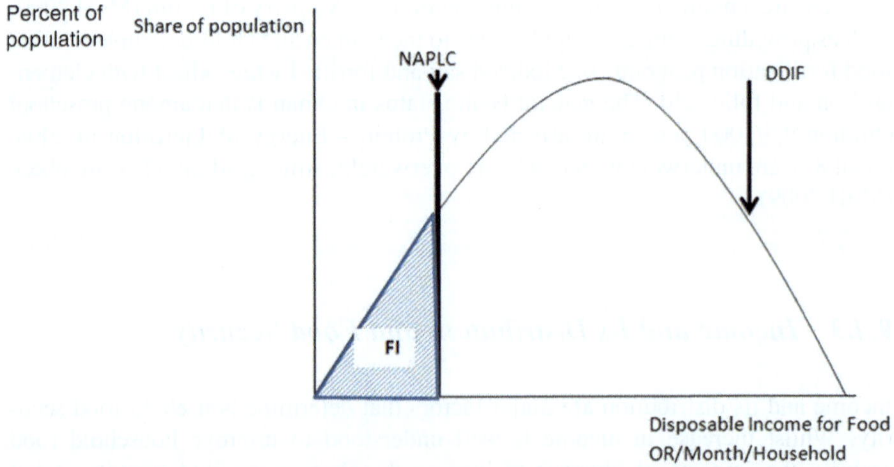

Fig. 9.1 Conceptual exposition of the analytical method (Notes: *NAPLC* is the food insecurity threshold estimated as Nutritionally Adequate and Preferred Least Cost Diet, *DDIF* is Distribution of Disposable Income for Food derived from income distribution data of Oman and by considering the proportion of income spent on food by different income levels, *FI* area is a measure of food insecurity as households falling below the food security threshold (NAPLC)) (Source: Adapted from Naiken 2003)

9.5.1 Food Security Measures

Food security measurement requires measurement of the population below a given food security threshold, given the distribution of household disposable income for food. Mousvi (2006) has adapted Foster-Greer-Thorbecke (FGT) measures (Foster et al. 1984) to examine food security impacts of rice market liberalization in Iran. Orewa and Iyangbe (2009) have also adopted FGT measures to profile food security among rural low income urban dwellers in Nigeria, given the food price surge in since 2007. FGT measures have been adapted to measure food security in this study. FGT measure for food security of a population given Household Disposal Income for Food (HDIF) distribution and a food security threshold (NAPLC) is as follows.

$$F_{\propto} = \int_{0}^{q} \left(\frac{s-y}{s}\right)^{\propto} dy \tag{9.1}$$

Equation 9.1 in discrete terms is as Eq. 9.2

$$F_{\propto} = \frac{1}{N} \sum_{i=1}^{q} \left(\frac{s-y_i}{s}\right)^{\propto} \tag{9.2}$$

Where:

F_α is food security index for $\alpha = 0$, 1 or >1
α is a sensitivity parameter
N is the population size
s is food security threshold, disposable income level below which the household is food insecure (NAPLC is used in this study)
y_i is (HDIF) disposable income for food of the ith household
q is number of households $y < s$ (food insecure).

Following are alternative food security measures that can be derived from the Eq. 9.2 by varying α the sensitivity parameter.

1. **when F $\alpha = 0$; Head Count Index of Food Insecurity (F_0)**
   When $\alpha = 0$ Eq. 9.2 will be as Eq. 9.3

$$F_0 = \frac{q}{N} \tag{9.3}$$

F_0 is referred to as the Head Count Index of Food Insecurity as it is the ratio between the number of the people who are food insecure ($y < s$) over the total population of people (N), given the HDIF. F_0 does not measure the intensity of food insecurity i.e. how insecure are those under food insecurity. F_0 considers those with y very close to s, as well as those with y largely divergent from s are equally weighed. F_0 would not recognize as important; if food insecure persons become further insecure or even if a food insecure person become better-off yet below s. It considers a discrete change from food insecure to secure as changes in y between s. By use of F_0 measurement of food security, strategies that improve food security of those close to s would be considered most effective, whilst in a normative sense improving the food security of those further from s is more ethical.

2. **F when $\alpha = 1$; Food Security Gap Index (F_1)**

   When $\alpha = 1$ Eq. 9.2 becomes as Eq. 9.4, which defines the Food Security Gap Index F_1.

$$F_1 = \frac{1}{N}\sum_{i=1}^{q}\left(\frac{s - y_i}{s}\right) \tag{9.4}$$

F_1 is a measure of amount of income that is required to bring all household that are food insecure to s (food security threshold), weighted by population size and s. It is interpreted as the required increase in disposable income for food of an average household to eliminate food insecurity or how much income would have to be transferred to the food insecure households to make them food secure. This is the sum of food security gaps ($s - y_i$). The sum of all food security gaps over the head count of $y < s$ is the average increase in the disposal income that is required to raise the HDIF of those who are food insecure to NAPLC (food security). This food security gap per individual food insecure (AFSG) can be estimated by Eq. 9.5.

$$AFSG = \frac{F_1 \times s}{F_0} \tag{9.5}$$

Thus AFSG is the estimate by which the average disposal income on food should be increased to reach food security of a household that is food insecure. Above measures are calculated using the ADePT: Software Platform for Automated Economic Analysis (World Bank 2010b). ADePT is a software that has been developed to automate and standardize the production of analytical reports. ADePT as described by the developer uses micro-level data from various types of surveys, such as Household Budget Surveys, Demographic and Health Surveys and Labor Force surveys to produce rich sets of tables and graphs for a particular area of economic research.

9.5.2 Decomposition of the Influence of Income and Its Distribution on Food Security

Ravallion and Datt (1991) and Datt (1998) provide a detailed explanation on computational tools for poverty measurement and on the rationale and methodology of decomposition of income growth and its redistribution as factors causing poverty

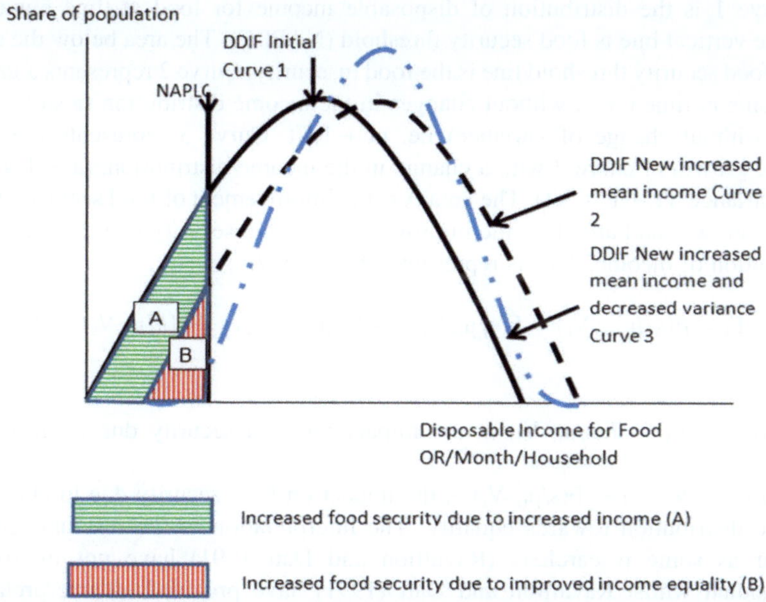

Fig. 9.2 Conceptual illustration of the impact of increase in income and improved income equality on food security (Notes: *NAPLC* is the food insecurity threshold estimated as Nutritionally Adequate and Preferred Least Cost Diet, *DDIF* is Distribution of Disposable Income for Food derived from income distribution data of Oman and by considering the proportion of income spent on food by different income levels) (Source: Adapted from Datt et al. undated)

with applications to Brazil and India in the 1980s. In India during the 1980s improved income distribution has countervailed the adverse impacts of droughts on food security. In Brazil the worsening income distribution has had adverse effects on poverty. Datt (1998) explains methodology to analyze poverty given secondary summary data that are commonly published on income distribution by simulating Lorenz curves on income distribution. The methodology suggested by Ravillion and Datt (1998) is used in this study with adaptation to food security given distribution of household disposable income for food. Growth in per capita income and more equal distribution of income are expected to improve food security (Fig. 9.2).

Food security measure at a time t (F_t) can be represented by Eq. 9.6.

$$F_t = F(s/\mu_t, V_t) \tag{9.6}$$

Where:

s is a food security threshold,
μ_t is the mean of the distribution of disposable income for food,
V_t is the variance of the distribution of disposable income for food.

Curve 1 is the distribution of disposable income for food at time t with μt, Vt. The vertical line is food security threshold (NAPLC). The area below the curve 1 and food security threshold line is the food insecurity. Curve 2 represents a growth in income at time t + 1, without changes in the income distribution (a shift of the mean without change of variance) i.e. μt + 1,Vt. Curve 3 represents the same income growth of curve 2 with a change in the income distribution, μt + 1,Vt + 1 (less variance Vt + 1 < Vt). The area A is the improvement of food security due to income growth and area B is the improvement in food security due to more equal distribution of income. This is represented by Eq. 9.7.

$$F_{t+1} - F_t = F\left(s/\mu_{t+1}, V_t\right) - F(s/\mu_t, V_t) + F\left(s/\mu_{t+1}, V_{t+1}\right) - F(s/\mu_t, V_t) + \text{Residual} \tag{9.7}$$

$F(s/\mu_{t+1}, V_t) - F(s/\mu_t, V_t)$ is the impact on food security due to growth in income.

$F(s/\mu_{t+1}, V_{t+1}) - F(s/\mu_t, V_t)$ is the impact on food security due to change in income distribution towards equality. The interpretation of the residual remains unclear as some researchers (Ravallion and Datt 1991) have not interpreted the residual whilst Ravallion and Datt (1991) have provided an interpretation. This study does not consider the residual in the interpretation. Above measures were calculated using the ADePT: Software Platform for Automated Economic Analysis (World Bank 2010b).

9.5.3 Distribution of Disposable Income for Food

The distribution of income for Oman was simulated through Monte-Carlo method using secondary data provided by MNE (2001, 2010a). Proportion of income spent for food by different income groups was calculated from primary data of a Food Demand Survey (Mbaga and Kotagama 2010). The population distribution of income was converted to population distribution of Household Disposable Income for Food (HDIF) per month by using the estimates of proportion of income spent for food.

9.6 Analysis and Discussion

9.6.1 Food Price Surge in Oman

The world food prices surged substantially in year 2008. The impact of the surge in world food prices on Oman's food prices is that the Consumer Price Index of all food in Oman has increased by 21.60 % in year 2008 from 2003. The prices of all

Table 9.2 Estimates of Nutritionally Adequate Preferred Least Cost Diet over years in Oman

Year	2003	2004	2005	2006	2007	2008
NAPLC OR/Month/Household	163.12	164.65	168.43	169.89	180.06	213.09

Table 9.3 Income distribution in Oman

	Year 1999/2000	Year 2007/2008
Income (OR/Month/Household)	% Households'	% Households'
Less than 100	8.20	3.00
100–199	12.30	4.70
200–299	13.30	7.60
300–399	12.40	9.00
400–499	10.40	9.50
500–599	8.80	6.50
600–699	5.90	5.70
700 and more	28.70	54.00
Average income (OR/Month/Household)	638.00	913.00

Source: MNE (2001, 2010a)

food groups have increased, but the most significant increases (more than 40 %) occur in the food groups; oil and fats, cereals and cereal products, followed by (more than 20 %) milk and milk products, eggs, fish, meat and poultry. The Value of Nutritionally Adequate and Preferred Least Cost Diet (NAPLC) as estimated by Al Jabri and Naser (2011) is given in Table 9.2. It is apparent that given the stability of food prices from 2003 to 2006 NAPLC has changed only by about 6 OR per month per household. However since 2006 with increased food prices, the value of NAPLC diet has drastically increased by 43 OR per month per household by 2008, i.e. an increase of 26 %.

9.6.2 Income Distribution in Oman

Table 9.3 gives the data on income distribution in Oman during years 2000 and 2008 (MNE 2001, 2010a) and the respective Lorenz curves are given in Fig. 9.3. The Lorenz curves and the Gini coefficient were estimated using Adept software (World Bank 2010b). Gini coefficient is a measure of inequality. The coefficient varies between 0, which reflects complete equality and 1, which indicates complete inequality. It is clearly evident that the income distribution has significantly improved with the Gini coefficient changing from 46.49 to 36.35 from year 2000 to 2008 in Oman along with average household income changing from 638.000 to 913.00 OR per month per household. Gini coefficient for Oman as reported by MNE (2010a) is 33.00.

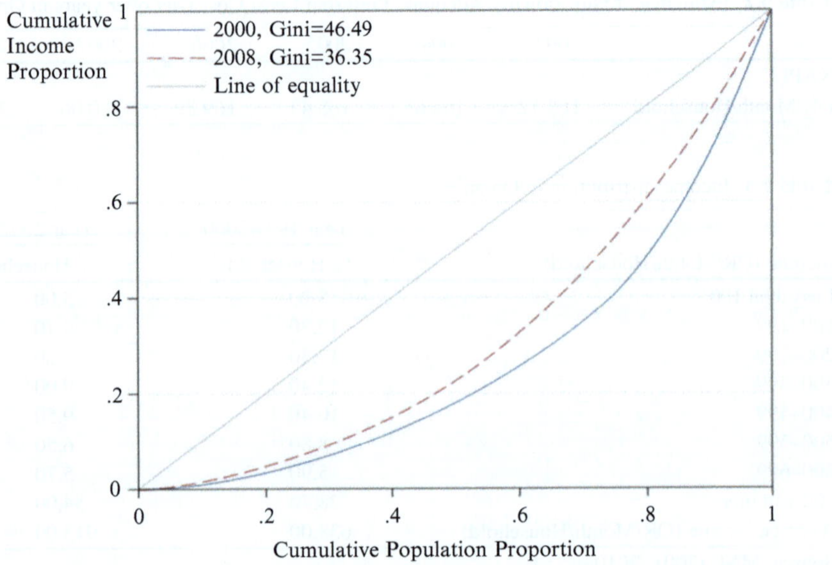

Fig. 9.3 The Lorenz curves for income distribution for years 2000 and 2008

The Monte-Carlo simulation was used to simulate the income distribution considering the statistics of income distribution of years 2000 and 2008 (Table 9.3) for a population size of 1,000. The descriptive statistics of the simulated data (Table 9.4) in comparison to the actual parameter, such as the average income confirms the robustness of the simulation. The actual average monthly household income for years 2000 and 2008 as reported by MNE (2001, 2010b) are 638.00 OR and 913.00 OR, respectively and that of the simulated data are 641.65 OR and 907.54 OR respectively.

9.6.3 Household Disposable Income for Food by Income Levels

The proportion of expense on food diminishes with increased household income (Engel's Law). Table 9.5 gives the estimates of proportion of income spent on food by different income levels by Omani households (Mbaga and Kotagama 2010). On the average an Omani household spends 21 % of income on food. According to MNE (2001, 2010b) the percent of income spent on food is 32.4 % and 31.1 % in years 2000 and 2008 respectively. The estimates of percent income spent on food (Table 9.5) were used to estimate the population distribution of household disposable expense on food by households with different income levels.

Table 9.4 Descriptive statistics of the simulated income distribution data

Statistics	Year 2007/2008	Year 1999/2000
Mean	907.54	641.65
Standard error	18.29	17.95
Median	788.96	438.38
Mode	319.21	381.44
Standard deviation	578.22	567.63
Sample variance	334,342.08	322,205.51
Kurtosis	−1.14	0.21
Skewness	0.37	1.17
Range	2,026.51	2,120.86
Minimum	2.95	3.05
Maximum	2,029.46	2,123.91
Count	1,000	1,000

Table 9.5 Percent of income spent on food with increasing household income

Income (OR/Month/Household)	% Expense on food of total household income
Less than 100	78.00
100–199	78.00
200–299	72.00
300–399	66.00
400–499	54.00
500–599	42.00
600–699	33.00
700 more	24.00

Source: Mbaga and Kotagama (2010)

9.6.4 Food Security Measures

FGT food security measures were estimated based on the population distribution of disposable expenditure on food by Omani households and the NAPLC estimates for years 2003 base year and 2008 when prices increased, were used. The ADePT software (World Bank 2010a, b) was used for the estimation of FGT food security measures. Table 9.6 gives scenario analysis (of price levels and income distributions of years 2003 and 2008) on F_0 and F_1 food insecurity measures. Higher values of these measures indicated reduced food security.

Food Security Head Count (F_0), i.e. the percentage of population unable to access NAPLC diet in year 2003 (2003 prices and 2003 income and its distribution) was 24.0 %. Due to the increase food prices in year 2008, given the 2008 income and its distribution F_0 is 29.3 %. This is a drop in food security by 5.3 % (24.0–29.3) due to food price increase despite the increase in income and better distribution of income between years 2003 and 2008.

Hypothetically had the prices of 2003 prevailed without increases?, F_0 would have decreased to 9.7 % with the 2008 income and its distribution. Thus food

Table 9.6 Measures of food security revealing the impact of price increases, income and its distribution on food security in Oman

	Food security			Food security gap (F_1)		
	Headcounts rate (F_0)					
	Income distribution			Income distribution		
	2003	2008	Change	2003	2008	Change
2008 prices	42.8	29.3	−13.5	15.2	6.9	−8.3
2003 prices	24.0	9.7	−14.3	10.0	3.7	−6.3

Note: Changes shown between years 2003 and 2008

Table 9.7 The cost of alleviating food insecurity

Parameter	Data and estimate
1. NAPLC 2008 (OR per month per household)	213.00
2. Estimated F0	0.29
3. Estimated F1	0.07
4. Average food insecurity gap (OR per month per household) (Eq. 9.5)	50.68
5. Oman's population (Million)	2.30
6. Number of household (Million) (5)/8) 8 members per household	0.28
7. Number food insecure [F0X(5)]	0.67
8. Total food gap [(7) × (4) × 12 months] (Million OR/Year)	174.84
8. GDP 2008 market prices (Million OR)	23,185.10
9. Food insecurity gap/GDP as %	0.75

security would have improved by nearly 14.3 %. Had the income and income distribution not changed and remained as 2003?, the food price increase would have led to F_0 as 42.8 %. Hence the change in income and its distribution between 2003 and 2008 has mitigated the impact of food prices on food security by 13.5 % (42.8–29.3 %).

Food Security Gap (F_1) changed by similar magnitudes and directions of change as F_0 during 2003 and 2008 (Table 9.6). The average food security gap was estimated as 50.680 OR per month per household given the estimates of F_0, F_1, NAPLC of year 2008 using Eq. 9.5 (Table 9.7). Thus an improvement of the income of the food insecure households by improvements in income through livelihood improving investment or targeted government transfers of 50.680 OR per month per household would eliminate food insecurity given food prices of year 2008. The total cost of such an income transfer would be about 0.75 % of the GDP of year 2008 as estimated (Table 9.7).

9.6.5 Decomposition of the Impact of Growth of Income and It Distribution on Food Security

It is apparent from above analysis that increase in food prices has decreased food security. However the increase in food prices is a recent phenomenon. Oman has been adopting policies to improve food security, through increases in per capita

Table 9.8 Decomposition of food security changes due to income growth and its redistribution

	Income distribution		Change in incidence of food security			
	2003	2008	Actual change	Growth	Redistribution	Interaction
Food security headcount rate (F_0)	24.02	9.70	−14.32	−6.31	−6.92	−1.09

Note: Base food prices of 2003 are considered

income and improving income distribution. The impact of these two factors on improving food security was analyzed. NAPLC on base prices of year 2003 was considered in the analysis to control the impact of food prices on food security. Table 9.8 gives the results of the analysis.

It is evident from the analysis reported in Table 9.8 that if the food prices had not increased F_0 would have decreased by 14.32 % to bring down food insecurity to 9.7 %. As the results reveal this improvement in food security has been caused by almost equally through growth income (6.31 %) and improvements in the distribution of income (6.92 %). Thus the Government of Oman has in the past implemented policies that have equal growth and equity oriented with a significant impact in reducing food insecurity.

9.7 Conclusions

It was found that the Consumer Price Index of all food in Oman increased by 21.60 % between years 2003 and 2008. The significant increases in prices of more than 40 % are of food groups; oil and fats, cereals and products, following with price increases of more than 20 % in milk and milk products, eggs, fish, meat and poultry. With increased food prices from 2006 to 2008 NAPLC has drastically increased by 43.198 OR per month per household. The analyses found that the income distribution has significantly improved with the Gini coefficient changing from 46.49 to 36.35 from year 2000 to 2008 in Oman along with average household income changing from 638 to 913 OR per month per household in Oman, during the same period.

Given above changes in food prices, income and its distribution, it was estimated that the Food Security Head Count (F_0), i.e. the percentage of population unable to access NAPLC diet in year 2003 (2003 prices and 2003 income distribution) was 24.0 %. Had the prices of 2003 prevailed?, F_0 would have decreased to 9.7 % with the 2008 income and its distribution. Thus food security would have improved by nearly 14.3 %. However due to the increase food prices in year 2008, given the 2008 income and its distribution, F_0 had increased to 29.3 %. The average food security gap i.e. amount of income that is required to bring all household that are food insecure to the food security threshold NAPLC, was estimated at 50.680 OR per month per household in year 2008. The total cost of an income transfer to cover the food security gap was estimated to be about 0.75 % of the GDP of year 2008. The results of the decomposition analysis revealed that food security would have improved by 14.3 % bring down food insecurity to 9.7 % had the food prices not increased?. The improvement in food security has been caused by almost equally through growth income (6.31 %) and improvements in the distribution of income (6.92 %). Thus it is indicated that the Government of Oman has in the past implemented policies that have been equally growth and equity oriented with a significant impact in reducing food insecurity.

The analysis indicates that food security in Oman could have drastically improved due to policies that have increased the per capita income and improvements in the distribution of income towards equality up to 2008. However the drastic increase in food prices since 2008 has increased food insecurity. Short term interventions by the government and assistance to vulnerable low income households would alleviate the situation. Continuing the implementation of egalitarian economic policies on investments in regional rural development, education, health, etc. will revert and further improve the food security situation in the Sultanate of Oman. Since 2008 the government of Oman has adopted a mix of such policies of improving social assistance to poor households, managing food prices, building food stocking capacity and drafting a strategy to assure long-term food security.

References

Adamowicz M (1988) Grain production and food security in Arab countries. Agric Econ 2:39–56

Al Jabri SAN, Naser SA (2011) Economic analysis of household food security in the Sultanate of Oman. Unpublished MSc thesis, Department of Natural Resource Economics, College of Agricultural and Marine Sciences, Sultan Qaboos University, Sultanate of Oman

Alasfoor D, Rajab H, Hadadd R, Al Rassasi B (2009) Food based dietary guidelines. Ministry of Health, Sultanate of Oman

Datt G (1998) Computational tools for poverty measurement and analysis. FCND discussion paper, no. 50, International Food Policy Research Institute, 2033 K Street, N.W. Washington, DC 2006

Datt G, Ramdas K, van de Mensbrugghe D, Walker T, Wodon Q (undated) Predicting the effects of aggregate growth on poverty. World Bank. http://siteresources.worldbank.org/INTPSIA/Resources/490023-1121114603600/12934_chapter10.pdf

Dawe D, Morales-Opazo C (2009) How much did developing country domestic staple food prices increase during the world food crisis? How much have they decline? ESA working paper no. 09–09. Agricultural Development Economics Division, Food and Agriculture Organization, United Nations

El-Sherbani A, Sinha R (1978) Arab agriculture-prospects for self sufficiency. Food Policy 3(2):84–94

FAO (2003) Trade reforms and food security, conceptualizing the linkages, Chapter 2. In: Food security: concept and measurement. Economic and Social Development Department, Food and Agriculture Organization (FAO). http://www.fao.org/docrep/005/Y4671E/y4671e00.htm

FAO (2008) Soaring food prices: facts, perspectives, impacts and actions required. Paper presented at high-level conference on world food security: the challenges of climate change and bioenergy. www.fao.org/foodclimate

Foster J, Greer J, Thorbecke E (1984) A class of decomposable poverty measures. Econom J 52(3):761–766

Gabbert S, Weikard HP (2005) The link between poverty and undernourishment: uncovering a methodological flaw. Rev Agric Econ 27(1):150–155

Gilbert CL, Morgan CW (2010) Food price volatility. Philos Trans R Soc 365:3023–3034

Leathers HD, Foster P (2004) The world food problem: tackling the causes of under nutrition in the Third World. Lynne Rienner Publishers, Boulder/London

Malthus T (1798) An essay on the principle of population, with a summary view (1830), and Introduction by Professor Anthony Flew. Penguin Classics. ISBN 0-14-043206

Mbaga M, Kotagama H (2010) Food demand survey indicative demographic trends. Unpublished, Department of Natural Resource Economics, College of Agricultural and Marine Sciences, Sultan Qaboos University, Sultanate of Oman

MNE (2001) A report about household expenditure and income survey results from 20/5/1999 to 19/5/2000. Ministry of National Economy (MNE), Directorate General of Social Statistics, Sultanate of Oman

MNE (2010a) Statistical year book 2010 Sultanate of Oman. Ministry of National Economy (MNE), Directorate General of Social Statistics, Sultanate of Oman

MNE (2010b) The important result of household expenditure and income survey results from 20/5/2007 to 19/5/2008. Ministry of National Economy (MNE), Directorate General of Social Statistics, Sultanate of Oman

MoH (2009) National nutrition surveillance, national nutrition surveillance report. Ministry of Health (MoH), Department of Nutrition, Sultanate of Oman

Mousvi SN (2006) The relationship between rice market liberalization and food security in Iran. Poster paper, International Association of Agricultural Economists conference, Gold Coast, Australia, 12–18 August

Naiken L (2003) Keynote paper: FAO methodology for estimating the prevalence of undernourishment. FAO Corporate Document Repository, http://www.fao.org/docrep/005/y4249e06.htm

Orewa SI, Iyangbe CO (2009) The food insecurity profile among the rural and low income urban dwellers in Nigeria. Am Eurasin J Sci Res 4(4):302–307

Pinstrup-Anderson P, Caicedo E (1978) The potential impact of changes in income distribution on food demand and human nutrition. Am J Agric Econ 60:402–415

Ravallion M, Datt G (1991) Growth and redistribution components of changes in poverty measures: decomposition with applications to Brazil and India in the 1980s. J Dev Econ 38(2):275–295

Senauer B, Sur M (2001) Ending global hunger in the 21st century: projections of the number of food insecure people. Rev Agric Econ 23(1):68–81

Timmer CP (2000) The macro dimensions of food security: economic growth, equitable distribution and food price stability. Food Policy 25:283–295

World Bank (2000) A better world for all: progress towards the international development goals. World Bank, Washington, DC, 20433

World Bank (2009) Improving food security in Arab Countries. World Bank, Washington, DC, 20433

World Bank (2010a) Guidance for responses from the human development sector to rising food and fuel prices. World Bank, Washington, DC, 20433

World Bank (2010b) ADePT website: www.worldbank.org/adept

Chapter 10
Agricultural Research and Development and Food Security in the Gulf Region

Huzur Keskin

Abstract This chapter attempts to achieve two objectives: the first consists of mapping out the current Research and Development (R&D) on agriculture and food security in the Gulf Cooperation Council (GCC) countries and the world; and the second is to advocate the case for urgent need of further and immediate investment in R&D in this sector. These two objectives are framed within a wider discussion of the growing vulnerability of the Gulf region in terms of food and agriculture. Identifying aspects of this vulnerability along with adverse side effects of adopting outsourcing food security policies would further expose the urgency of R&D in this area, and points at issues and agendas that should be granted priority. The multiplicity of pressures faced by the GCC in the field of meeting local demands for food provision generate a high level of uncertainty to be left without close scientific research and problem-solving, both represent a prerequisite for making right decisions and draft effective strategies.

Keywords GCC • R&D • Agriculture • Food security • Public sector

10.1 Introduction – An Overview of Global R&D on Agriculture

According to Bardhan (1996) the world's population is estimated to be around 7.0 billion and to become over 11 billion by the year 2050. The global food security shall crave at least a doubling or more of food production by the year 2050 to meet the demand of global population almost 90 % of whom will live in developing

H. Keskin (✉)
Finance, International Investment and Affiliates Manager, Hatir 20,
PO Box 06700, Cankaya, Ankara, Turkey
e-mail: k.huzur@gmail.com

S.A. Shahid and M. Ahmed (eds.), *Environmental Cost and Face of Agriculture in the Gulf Cooperation Council Countries: Fostering Agriculture in the Context of Climate Change*,
DOI 10.1007/978-3-319-05768-2_10,

countries. Now, 40 % of population is living in developing countries and even with today's population there are inadequate food supplies and abject poverty.

To survive the agricultural productivity, research and technological preferences have to continue and developed for farmers and the people at large. The global economic expansion cannot be thought of without agricultural growth which is the main locomotive sector now and in future. To have food security matching rapid global population growth, agriculture must be accorded high priority.

Both the public and private sectors in developed countries have benefitted from agricultural R&D investments whereas developing countries have generally relied on public sector support from national programs or international organizations such as the international centers of the Consultative Group on International Agricultural Research (CGIAR) (James 1996). In both industrial and developing countries, there will be a need to make not only national or international partnerships but also public and private affiliations to meet the challenge of food security.

The general fiscal restraint causes more market-oriented view than a skeptical concept of social benefits. The recent agricultural R&D policies in developed and developing countries have introduced the same change in their priorities. The potential benefits from agricultural R&D with the future global demand bring the "non-traditional" interest groups on the market as players (McMillan et al. 2001). These new players help to have new fundamental biological sciences and scientists with supporting information technologies into genetics of agriculture for worldwide.

The progress of modern biotechnology in agriculture brings a new understanding to share the knowledge of agricultural information and methods, yet limited by the regulations of "intellectual property rights (IPR)". With IPR regime, the new technologies have not been transferred freely anymore. It means that countries and/or companies will be dependent on the others who already acquire developed technologies.

The well-known truth is that biotech companies in the world are mostly located in the rich countries which are no longer interested in the simple productivity enhancement. The predominant concern in poor countries is about food security while rich countries focus on certain attributes of food, food production system and other market-oriented issues. The differences of priorities between rich and poor countries will affect progress in global agriculture sector in the near future (Braun 2008). This question should be managed seriously and carefully. By the end of the last century, the public and private sector's roles shifted on agricultural R&D. Policy designers, decision makers of resource allocation, developers and scientists are now facing a new era with regard to the agricultural R&D.

10.2 Public Sector Investments in Agricultural R&D

Over the past two decades of the twentieth century, worldwide public investments in agricultural research increased by 51 % in inflation-adjusted terms. Remarkably, during the 1990s, developing countries made more of the world's public agricultural R&D than the developed countries (Pardey et al. 2006a). In the past few years, changes in the governments' role in the agricultural R&D have brought some new approaches (Pardey et al. 2006a), including:

- Advancing IPP (intellectual property protection);
- Establishing levy compromise;
- Tax concessions.

10.2.1 Historical Development of Public Agricultural Research

The first agricultural research institutions were formalized in the nineteenth century in Europe. The practice of providing public funds aimed to support national agricultural R&D agencies staffed with professional scientists (Grantham 1984). Consequently, scientific agricultural foundations had become fully developed and the progress created a second steady wave of addition in public charge on the agricultural R&D at the beginning of the twentieth century (Grantham 1984).

Modern biotechnology which has been based on DNA techniques has changed the practice of agricultural R&D and brought with it the practice of intellectual property protection (IPP). The legislation of IPP had caused the balance of agricultural research between locally provided international traded R&D products and services. After the World War II, with the improvements of science and technology, in the developed countries beside the public sector, the private sector started to become interested in the agricultural R&D.

10.2.2 Global Perspective of Public Agricultural R&D

Developed countries in 1960s had almost two-thirds of the total public sector investments in the global agricultural R&D. In 1990s, the picture changed and the developing countries invested more than industrial countries. In 2000s, percentage of national agricultural GDP of developing countries was still higher than that of the developed countries' R&D intensity (the percentage of national agricultural GDP) (Table 10.1).

While there was a declining pattern in growth rates of agricultural R&D in industrial countries in 1980s, developing countries, China included with its large portion of total investments, have started to have booming growth rates in agricultural R&D in the same period (IFPRI 2008). The Asia and Pacific region has continued to have an ever-expanding share of the developing countries whereas the share of China and India constituted 40 % of total expenditure on agricultural R&D in developing world.

Table 10.1 Total public agricultural research expenditures by region, 1981, 1991 and 2000

Expenditures (million 2000 international dollars)	1981	1991	2000
Developing countries	6,904	9,459	12,819
Sub-Saharan Africa	1,196	1,365	1,461
China	1,049	1,733	3,150
Asia and Pacific	3,047	4,847	7,523
Latin America and the Caribbean	1,897	2,107	2,454
Middle East and North Africa	764	1,139	1,382
Developed countries	8,293	10,534	10,191
Total	15,197	19,992	23,010
Annual growth rates (percent per year)	**1981–1991**	**1991–2000**	**1981–2000**
Developing countries	3.04	2.90	3.14
Sub-Saharan Africa	1.25	0.82	0.99
China	4.76	5.04	4.86
Asia and Pacific	4.33	3.92	4.19
Latin America and the Caribbean	1.13	2.06	2.01
Middle East and North Africa	4.12	1.87	3.35
Developed countries	2.27	–0.58	1.10
Total	2.63	1.20	2.11

Source: Agricultural Science and Technology Indicators (ASTI) data underlying Pardey et al. (2006a, b)

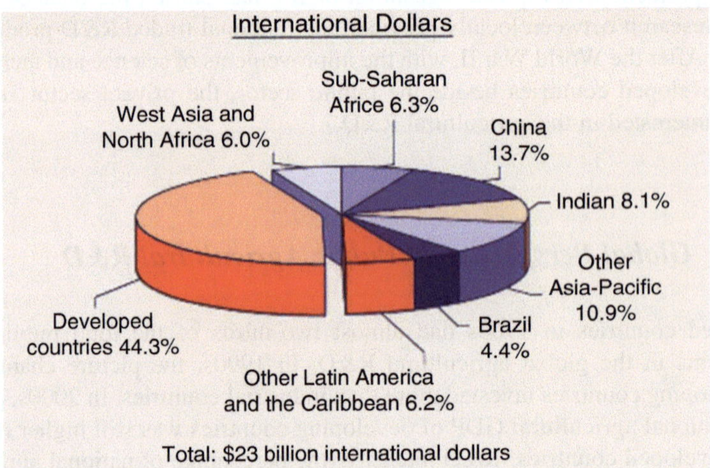

Fig. 10.1 Public agricultural researches spending in international dollars, 2000 (Source: Calculated by author based on data reported in Table 10.1)

The public agricultural R&D consists of different performers, including governments, higher education agencies, and non-profit institutions. There are important differences between regions in the structure of public agricultural research sector (Fig. 10.1).

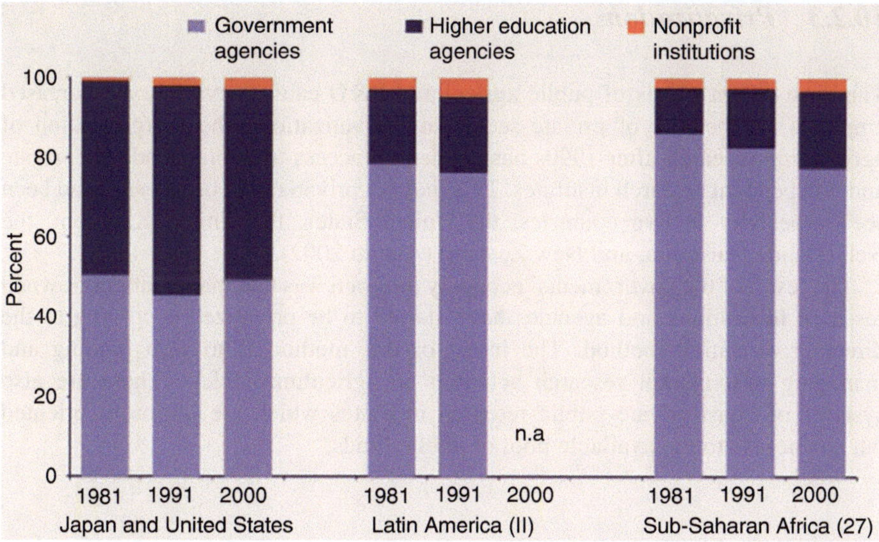

Fig. 10.2 The structure of public agricultural research sector (Source: Pardey et al. 2006b)

In the United States, public agricultural research is done mainly in State Agricultural Experiment Stations (SAES) located at colleges and administered federally. In Latin America, a larger share of public agricultural R&D is managed and done by government agencies. Similar situation exists in Sub-Saharan African countries (Fig. 10.2).

In addition to national public agricultural R&D activities, there are also international agricultural research agencies which are funded by national public institutions. The first of such ventures was Rockefeller Program initiated in 1943, followed by other international research institutions such as the International Wheat and Maize Research Centre (CIMMYT). Currently, the well-known and active international agricultural research centre is the Consultative Group on International Agricultural Research (CGIAR) which was established in 1971 (Alston et al. 1998).

The recent developments in public agricultural R&D policy could be summarized as follows: "*(a) trends towards using public funds to support more basic rather than more applied or 'near-market' research; (b) trends towards the joint funding of near-market research through the development of industry levies and other mechanisms; (c) revamped oversight and accountability mechanisms, and other changes in management of research resources; (d) measures for introducing competition among researchers to increase research productivity and as a means of allocating research resources; (e) measures to privatize, directly or indirectly, public agricultural research institutions; and (f) trends towards the rationalization of public agricultural research facilities (Alston et al.* 1998, 2000*)*".

10.2.3 Privatization

The most recent trends of public agricultural R&D cause inevitably an increased emphasis on the role of private sector and privatization. The reorganization of agricultural research after 1990s has opened up access to public funds for private and independent research institutes. This indirect privatization of research had been seen especially in five countries, the United States, the United Kingdom, the Netherlands, Australia, and New Zealand (Alemu 2002).

The result of the governments' economy program was that many publicly-owned research institutions and agencies have started to be privatized according to the direct privatization method. The intent of this method is to stop funding and managing near-market research activities in agricultural R&D. There are also systems of quasi-private public research institutes which are not-profit oriented but can access to the available pool of public funds.

10.2.4 Private Sector Investments in Agricultural R&D

Although agricultural innovation has mainly been an individual undertaken, individuals nevertheless cannot fully maintain reasonable returns from their research investments. For this reason, there have been collectively conceived and funded agricultural R&D by public. The public agricultural R&D warrants ensuring an adequate investment in research but private agricultural R&D has been continued to flourish since 1970s (Ahmed and Rustogi 1987).

10.2.5 Private Sector Agricultural Research

At present, there are huge private sector investments in agricultural R&D around the world, but with dramatic differences between developed and developing countries. In developed countries, almost 50 % of total agricultural R&D investments have been made by private sector but over the 90 % of total agricultural research expenditures are public investments in developing countries (Bentley and Tewari 1997) (Table 10.2).

The private sector main activities in agricultural R&D focus on the development, production and distribution of products and services dedicated to commercialization (James 1996). The private sector has continued to emphasize inventions that are amenable to intellectual property protection (IPP), patents, rights and others. Potential market size, the costs, communication and transportation infrastructure, size of farms, income and profit margins are very important issues for the private investments.

Table 10.2 Global public and private agricultural R&D investments in 2000

	Expenditures (million 2000 international dollars)			Share (percent)	
Region/country	Public	Private	Total	Public	Private
Asia–Pacific	7,523	663	8,186	91.9	8.1
Latin America and the Caribbean	2,454	124	2,578	95.2	4.8
Sub-Saharan Africa	1,461	26	1,486	98.3	1.7
Middle East and North Africa	1,382	50	1,432	96.5	3.5
Developing-country subtotal	12,819	862	13,682	93.7	6.3
High-income country subtotal	10,191	12,086	22,277	45.7	54.3
Total	23,010	12,948	35,958	64.0	36.0

Source: Agricultural Science and Technology Indicators (ASTI) initiative data and data presented in OECD (2005)

In agricultural R&D investments, the merging and acquisitions (M&A) were initially experienced in the 1980s and 1990s. The trend of M&A in agricultural researches has been ultimately resulting in very large companies dominating in the international market. This trend is caused by the long-term research investments which need to ensure competitiveness and effective global market (IMF 2008).

The private sector investments, especially in biotechnology in agricultural R&D are multidisciplinary including pharmaceuticals. The increasing population cause more need to the pharmaceutical products. For this reason, this sector could be profitable investment area for the private sector. In recent years, most private sector R&D investments are in biotechnology research because the most private sector investments can be subjected to a degree of confidentiality.

10.2.6 Private-Public Links of R&D Funding

By mid-1990s, one-third of total global agricultural R&D investments were provided by private sector (Fig. 10.3). Only 6.3 % of this private R&D investment took place in developing countries. It means that public funds are still the main source of support in this category of countries. The size of the accumulated stock of knowledge is the meaningful measure of a country's technological capacity and a better account of cross-country differences in agricultural productivity (Alston and Pardey 1996).

Having enough funding for agricultural R&D come first, but putting in place the stock of knowledge and accumulate practices is equally necessary. For this reason, the most important incentive for collaboration between public and private sectors in agricultural R&D is to use the limited global resources in more effective way with the collective knowledge.

Governments do not view for profit private sector activities as detrimental to the public good to use them in most effective way to achieve national and global goals. The collective goal must be building partnerships that use comparative advantages

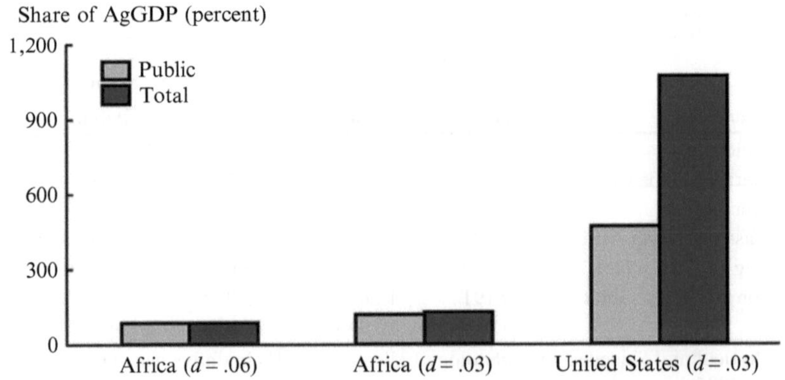

Fig. 10.3 African and American stocks of research knowledge in 1995 (Source: Alston and Pardey 1996)

of public and private sectors to get mutual targets (OECD and FAO 2008). Governments and private sector must take necessary and urgent steps to have a partnership in order to encourage impending challenge of global food security.

10.3 Food Security in the Gulf Region and Arab Investments in Agricultural R&D

There was a sharp rise in demand and prices of agricultural commodities and food in 2007 and 2008 which was a cause for concern about food security, malnutrition and increased levels of poverty. Food security requires a global response, involving governments, international agencies, non-governmental organizations, civil society and global private sectors. The increase of population, income growth, biofuel demand and other structural factors contribute in keeping prices high in agricultural sector. Within this global context, the worrying fact in the Arab and Gulf region is that countries are heavily dependent on food imports and have always been vulnerable to the fluctuation of global food prices.

Both supply and demand are elastic in terms of food prices which can increase rapidly in global market. Although public support for agricultural research has decreased since 1990s, food price shocks have continued to occur and increase. Beside this, the global growth rates of yields of cereals have been slowing down between 1980 and 2005 (Fig. 10.4). This means that there is bottleneck which causes the instability of food security (Wright et al. 2006) (Fig. 10.5).

Transportation and logistics are important costs which affect the level of food prices, raising alarms almost everywhere. In this context, it cannot be said that oil-producing countries need not to be concerned about high food prices. Even when oil prices are high, enabling these countries to afford paying for highly priced food commodities, high import expenses remain very exhausting and challenging.

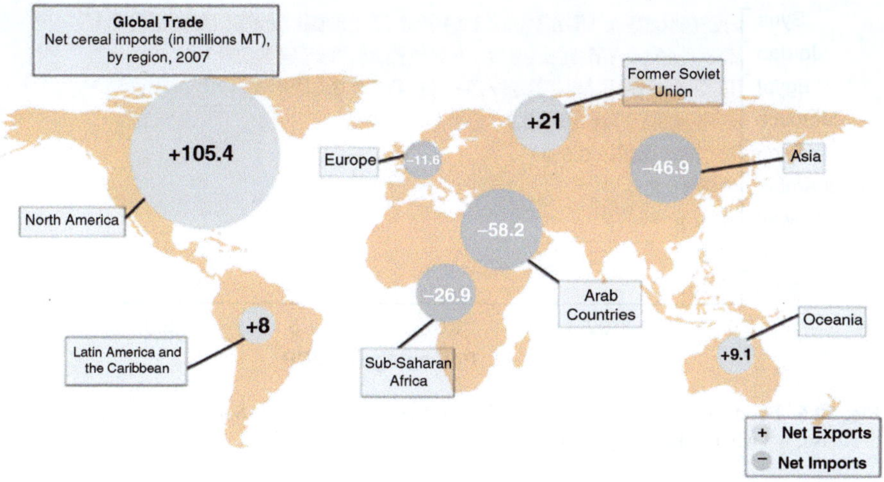

Fig. 10.4 Global trade of net cereal in 2007 (Source: FAO 2008)

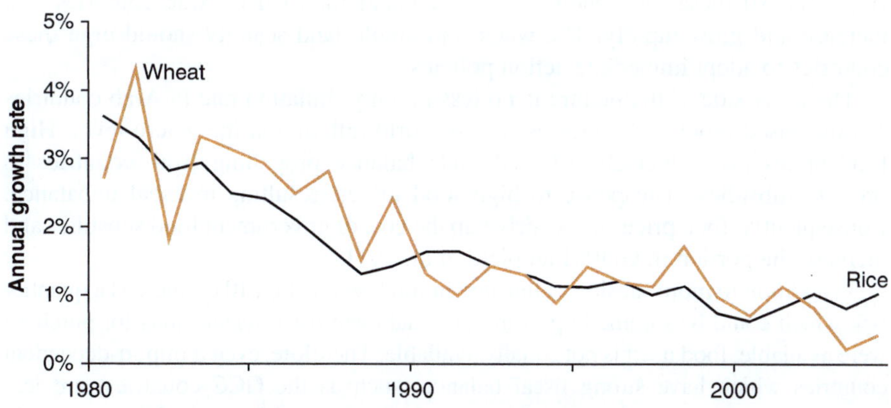

Fig. 10.5 Annual growth rates of major cereals in between 1980 and 2005 (Source: FAO 2008)

In fact, an increase in oil prices consequently leads to an increase in food import costs and logistics offsetting part of the surplus made by high oil prices.

In many Arab countries, there are population growth, urbanization and income growth all of which cause the increase of food demand. These countries also suffer from water and arable land scarcity with low supply difficulties, all of which create similarities among them more than with other regions elsewhere. The possible decoupling of oil and food commodity prices affect the fiscal balance of oil-producing countries and make them vulnerable to food prices and quantity shocks.

The population growth rate in the Arab countries is about 1.7 % while global growth rate is 1.1 % (World Bank 2008). Income growth rate is also bigger than the

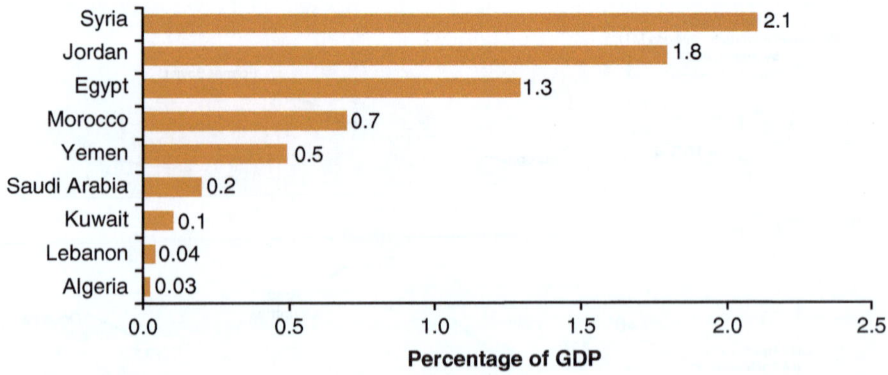

Fig. 10.6 Food subsidies in selective Arab countries GDP in 2007 (Source: World Bank online database, www.worldbank.org/data)

global rate which is 3–3.4 % in Arab countries. The urbanization rate in the same countries hovers around 3 % during 1990–2006 periods whereas global average rate is 2.2 %. All these rates show that the demand for food in Arab countries will increase and grow rapidly. The water and arable land scarcity should urge these countries to adopt immediate action policies.

The other side of the picture is no less gloomy. Inflation rate in Arab countries has increased more than twice as fast as world inflation at the macro level. High food prices also adversely affected trade balance propelling Arab countries to provide subsidies in response to high food prices, resulting in fiscal unbalance. Consequently, food price shocks drive up the cost of government food subsidies and increase the portion in GDP (Lampietti et al. 2011).

In addition to more subsidies allocated in food sector (Fig. 10.6), there is a quantity risk which could be summed up in the fact that even if sufficient funds for purchase were available, food itself is not equally available. Therefore, even if import-dependent countries which have strong fiscal balance (such as the GCC countries) are less vulnerable to price risk, they remain exposed to quantity risk; and in fact not a single Arab country can be protected from future food price and quantity shocks (Fig. 10.7).

According to the above, it may goes without saying that Arab governments must have preventive policies and programs such as trade policy, wage strategy and safety net programs (Table 10.3). All these policies and programs should aim to increase agricultural productivity and value per unit land and water. This requires investment in research and technology transfer.

The growing dependence on import makes the GCC countries highly vulnerable to external shocks and fluctuations in food prices. For this reason, policy makers in the region felt that in order to ensure food security, the most effective way for achieving sustainable agriculture was to rely on outsourcing. Therefore, the GCC nations have changed their agricultural policies from national-based and food self-sufficiency towards more direct investments in North-East Africa and South Asia in agriculture and food sectors.

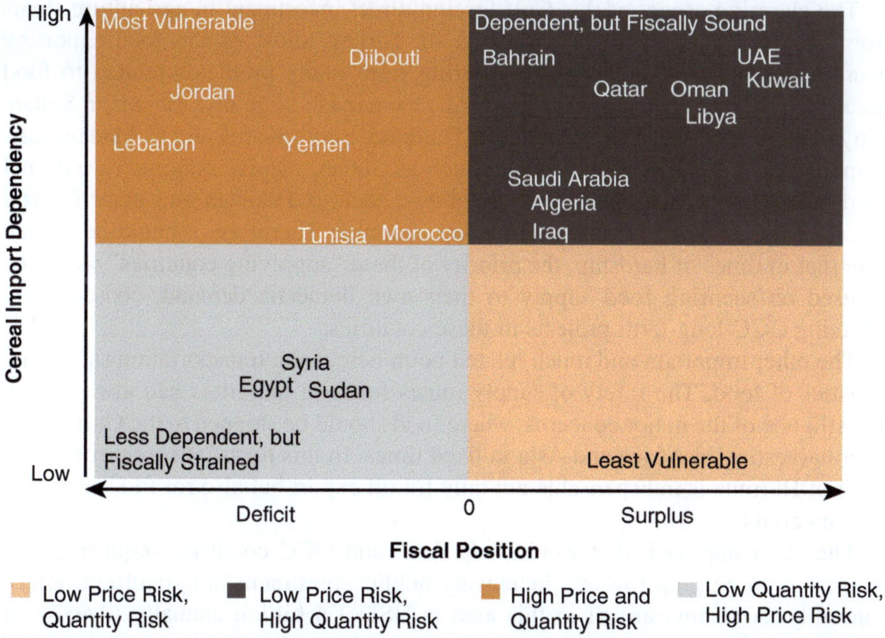

Fig. 10.7 Price and quantity vulnerability in Arab countries (Source: Lampietti et al. 2011)

Table 10.3 Economy-wide policies and existing social protection in Arab countries to protect the price shock

	Economy-wide policies				Existing social protection programs			
Country	Reduces taxes on foodgrains	Increase supply using foodgrain stocks	Export restrictions	Price controls/ consumer subsidies	Cash transfer	Food for work	Food ration/ stamp	School feeding
Egypt			√	√	√		√	
Morocco	√	√		√				√
Tunisia	√	√		√	√			
Yemen		√	√	√	√			
Lebanon	√			√				√
Syria	√	√		√	√		√	√
Jordan	√			√	√			√
West Bank and Gaza	√				√		√	√
Iraq	√	√	√	√	√		√	
Djibouti	√			√		√		√

Source: World Bank online database, www.worldbank.org/data

The alarming aspect of this Gulf foreign direct investment in agriculture stems from the fact that, with the exception of Turkey, most of the food-exporting countries to the Gulf are already suffering from many problems related to food security. The United Arab Emirates (UAE) have made large investments in Sudan, Libya and in Ukraine; Saudi Arabia in Thailand, South Korea and in Madagascar. Some of these "supplying" countries such as Sudan, receive continuous external food aid from international agencies; others, such as Pakistan and other Central Asian countries, suffer from severe water shortage. Therefore, it could be argued here that in times of hardship, the priority of these "supplying countries" would be focused on securing food supply to their own domestic demand, consequently affecting GCC long-term projects in these countries.

The other important and much related point here is the transportation safety and logistics of food. The safety of supply routes for food importers had always been and still one of the major concerns, where food should be shipped to the Gulf region after harvesting in Africa and Asia in fixed times. In this regard, the security of the strait of Hurmuz is indispensable not only for oil export but also for food and other raw materials.

The other approach that would help Arab and GCC countries acquiring food security is to gear up towards increasing public investment in agricultural R&D. Currently, Arab investment in this area is USD 1.4 billion annually (0.66 % of agricultural GDP): which is higher than the average in developing country, around 0.53 %, but is still lower than the level in developed countries of 2 %. There must be some incentives in the area of agricultural research in order to attract the academic staff to do their research in this region (Pardey et al. 2006b).

The League of Arab States (LAS) and the United Nations Development Program (UNDP) recommendation to set up a regional R&D fund with a committed long term budget to capture returns to research from beneficial spillovers is a major consequence of underinvestment at the national level (Alston and Pardey 1996). The Arab Centre for the Studies of Arid Zones and Dry Lands (ACSAD) was established by LAS in 1968 with a mandate similar to ICARDA and covers all Arab countries to help those achieving shared objectives (Lampietti et al. 2011).

10.4 Towards a New Model for GCC to Improve Agricultural R&D and Enhance Food Security

The GCC countries need to have new technology and tailor-made solutions, resulting from more indigenous research and development activities in agriculture sector. R&D is considered, in one definition, as a "*creative work undertaken on a systematic basis in order to increase the stock of knowledge, including knowledge of man, culture and society, and the use of this stock of knowledge to devise new applications*" (Wright and Zilberman 1993). This is exactly what has become clearly pressing in the agriculture sector in the region.

Multinational R&D companies, organizations or parties could be brought together and into the region with its convenient sources and facilities in order to create an investment in new agricultural technologies. It is almost well acknowledged that the main challenge of R&D activities is funding. However, funding should not be the biggest obstacle in the GCC region. Most Gulf cities are in fact branded as international financial centers with the GCC countries enjoying enormous sovereign wealth funds, mostly invested outside the region, part of which could be used to have R&D activities to develop new agricultural technology.

In order to embark onto a new model in this strategic area for future of their countries, the GCC governments should bring not only foreign professional managers but also scientists and researchers. Yet, integral part of this endeavor should be building and investing in local capacity, training and producing a generation of young scientists and researchers in this sector. An ambitious prospect is to start rethinking the region not only as financial centre but also as an agricultural R&D hub of the world; within this perspective GCC countries will not only produce agricultural solutions for their own needs, but also exporting technology. The GCC countries, attractive and satisfactory for the professional managers at the present, could also become attractive for experts and scientists who have knowledge and technology.

10.5 Mission

One of the aims of GCC governments in this context could be to generate new technologies and productivity in agriculture sector. With rapid industrialization and urbanization in the future, policy makers should encourage research institutes to become financially self–supporting with their export of new R&D commodities and services not only in Gulf region but also to the world. The evolution of the organizational structure, institutional management and financing of the agricultural research system in the GCC countries will certainly explore options to promote future agricultural growth and food security and reduce poverty.

The new technologies should be released by the agricultural research system in GCC countries. One of the long-term targets should be the creation of momentum where future growth in agricultural production will rely on continued productivity improvements. Related country-level activities should deal with technology transfer issues, such as demonstration trials, farmer education and other extension-related work. There must be private agricultural research and development initiatives and must be encouraged to be more active and global market-oriented.

10.6 Needs and Structures

In order to have new and sustained momentum in agricultural technology and productivity aiming to foster food security in the region, one could think of four areas of fund-channeling: (1) traditional publicly funded and managed research

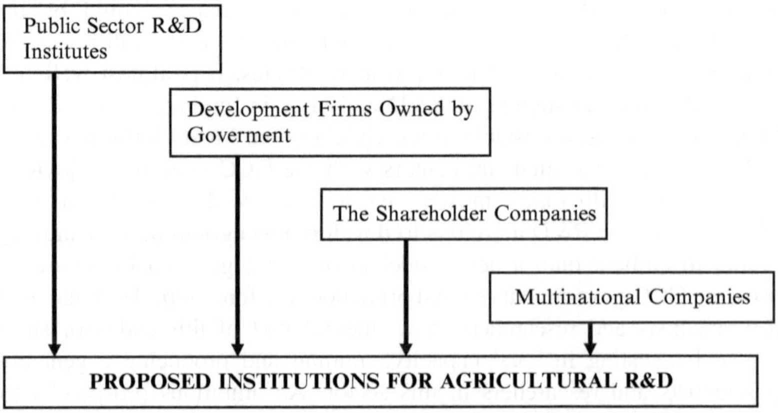

Fig. 10.8 Proposed Institutions for Agricultural R&D (Source: Designed by author)

institutions, (2) development firms owned by states, (3) shareholder companies owned by both public and private sectors, and (4) multinational companies and institutes (Fig. 10.8).

Public sector R&D institutes should be the backbone of the GCC countries' agricultural research system within the academies and agencies under the oversight of ministries of agriculture. The research efforts of various institutes must be under the administrative control to have relatively large size of share knowledge and scientific results to share whole country. These publicly owned institutes should be scientist incentive at the national level that focuses on pre-technology rather than site-specific research.

Development firms owned by government could be funded by government budgets but also should be encouraged to pursue commercialization of technology and the development of technology markets and rewarding individual scientist based on their efforts. The firms can contact directly the user of technology in not only local market but also global market. The main point for these government owned firms is that there must be low profit margin because they can use the fiscal local budget. They can be more market oriented in the global market but in national market, the services and new technology can be provided to the farmers free of charge.

The shareholder companies owned by both public and private sectors should be founded for specific purpose or market demand. They must be completely market oriented and should produce technologies and services as tailored make solutions. These companies should provide global competitive conditions and facilities to the researchers, scientists and professionals in the GCC region. These companies must be playing a magnet role for international agricultural R&D companies and scientist.

Multinational companies can bring global technology channeled from international expertise and companies that have led to rapid gains in productivity and output in GCC countries. These firms can play a larger role in the future and can be a kind of magnet like the shareholder companies. But the increase nature of property rights in the global market means that potential remains to be realized from the involvement of multinational companies. International corporations can prevent technology loss but when technology can be protected and market demand is high, fragmented retailing and wholesaling networks limit market penetration and it means that the profit margin of these multinational companies will be higher.

10.7 Strategy

In GCC countries, there is an investment climate in agricultural R&D and it can be strengthened in several ways. Academic and teaching institutions should be attracted by incentives and magnet facilities. Governments must develop innovative strategies that encourage the private sector from the national and international market to make investment in agricultural research with stronger intellectual property rights for improved varieties and other innovations.

The governments in the Gulf region could give preferential policy treatment, including tax exemptions, low-interest loans, new levy system and high quality of living conditions. The aim is to make the technology and knowledge mobilized and effective. Funding mechanism can direct for the local institutions and agencies, indirect for the national and international firms. There must be target to minimize the government funds' share slowly in the future. The agricultural R&D firms should be active by owned funds, but supporting farmers must be always governmental issue.

10.8 Conclusion

The aim of agro-investments of the GCC countries is to boost the production in agriculture in order to meet local demands, at the present and in the future. The GCC countries have to prepare and rehabilitate markets and infrastructure to create R&D activities in agricultural sector in the Gulf. They should have comprehensive development strategy and policies. They should have road map to attract R&D companies and scientists to the region.

The successful resolution of the food security in GCC countries should not be measured primarily by decrease in food prices. There must be a new boost in technological and policy innovation as essential measure to achieve food security and taking benefit from the global advantages. GCC countries should not be dependent neither on food supply nor on agricultural R&D technology. They must be able to produce their technology and even export their services and innovative products to the world.

References

Ahmed R, Rustogi N (1987) Marketing and price incentives in African and Asian countries: a comparison. In: Elz D (ed) Agriculture marketing strategy and pricing policy. World Bank, Washington, DC, p 58

Alemu AG (2002) Annotated bibliography on agriculture, food, rural development, and environment in Ethiopia. Vision Network for East Africa Report 2. Ethiopian Development Research Institute and International Food Policy Research Institute (IFPRI), Washington, DC, July, pp 65–66

Alston JM, Pardey PG (1996) Making science pay: the economics of agricultural R&D policy. American Enterprise Institute Press, Washington, DC, pp 22–73

Alston JM, Philip GP, Smith VH (1998) Financing agricultural R&D in rich countries: what's happening and why. Aust J Agric Res Econ 42(1):61–69

Alston JM, Chan-Kang C, Marra MC, Pardey PG, Wyatt TJ (2000) A meta-analysis of the rates of return to agricultural R&D: Ex pede Herculem. International Food Policy Research Institute (IFPRI) Research Report No. 113. Washington, DC, 111 pp

Bardhan P (1996) Research on poverty and development twenty years after redistribution with growth. In: Proceedings of the annual World Bank conference on development economics. World Bank, Washington, DC, 9 pp

Bentley WR, Tewari SC (1997) Evaluating benefits and cost for public investments in natural resource and environmental projects. In: Kerr John M et al (eds) Natural resource economics – theory and application in India. Oxford and IBH Publishing Co., New Delhi, 45 pp

Braun JV (2008) Food and financial crises – implications for agriculture and the poor. International Food Policy Research Institute (IFPRI), Washington, DC, 98 pp

FAO (Food and Agriculture Organization) (2008) The rise in crude oil prices stimulates ethanol-related demand for agricultural commodities. FAO, Rome

Grantham G (1984) The shifting locus of agricultural innovation in nineteenth-century Europe: the case of the agricultural experiment stations. Res Econ Hist 3:191–214

IFPRI (International Food Policy Research Institute) (2008) Agricultural science and technology indicators database. IFPRI, Washington, DC, 77 pp. Available at: http://www.asti.cgiar.org

IMF (International Monetary Fund) (2008) World economic and financial survey, pp 38–39. Online database. www.imf.org/external/pubs/ft/weo/2008/02/weodata/index.aspx. Accessed 25 Nov 2009

James C (1996) Agricultural research and development: the need for public-private sector partnerships. Consultative Group on International Agricultural Research (CGIAR), The World Bank, Washington, DC, 9 pp

Lampietti JA, Michaels S, Magnan N, McCalla AF, Saade M, Khouri N (2011) A strategic framework for improving food security in Arab countries. Published Online at Springerlink.com, 110 pp

McMillan MS, William AM (2001) An African growth trap: production technology and the time-consistency of agricultural taxation, R&D and investment. Purdue University, Medford, pp 202–203

OECD (Organization for Economic Cooperation and Development) and FAO (Food and Agriculture Organization) (2008) The OECD-FAO agricultural outlook, 2008–2017, pp 129–130. Online database. www.oecd.org/pages/0,3355,en_36774715_36775671_1_1_1_1_1,00.html. Accessed 15 Feb 2010

Pardey PG, Alston JM, Piggott RR (2006a) Agricultural R&D in the developing world: too little, too late? International Food Policy Research Institute (IFPRI), Washington, DC, pp 13–18

Pardey PG, Beintema N, Dehmer S, Wood S (2006b) Agricultural research – a growing global divide? International Food Policy Research Institute (IFPRI), Washington, DC, 6 pp

World Bank (2008) World Bank online database. Washington, DC. www.worldbank.org/data. Accessed 25 Nov 2009

Wright BD, Zilberman D (1993) Agricultural research structures in a changing world. In: Weaver RD (ed) US agricultural research: strategic challenges and options. Agricultural Research Institute, Bethesda, p 81

Wright BD, Pardey PG, Nottenburg C, Koo B (2006) Agricultural innovation: economic incentives and institutions. In: Evenson RE, Pingali P, Schultz TP (eds) Handbook of agricultural economics, vol 3. Elsevier, Amsterdam, 72 pp

Author Index

S.A. Shahid and M. Ahmed (eds.), *Environmental Cost and Face of Agriculture in the Gulf Cooperation Council Countries: Fostering Agriculture in the Context of Climate Change*, DOI 10.1007/978-3-319-05768-2, © Gulf Research Centre Cambridge 2014

Subject Index

S.A. Shahid and M. Ahmed (eds.), *Environmental Cost and Face of Agriculture in the Gulf Cooperation Council Countries: Fostering Agriculture in the Context of Climate Change*, DOI 10.1007/978-3-319-05768-2, © Gulf Research Centre Cambridge 2014

MIX
Papier aus verantwortungsvollen Quellen
Paper from responsible sources
FSC® C105338

If you have any concerns about our products,
you can contact us on
ProductSafety@springernature.com

In case Publisher is established outside the EU,
the EU authorized representative is:
Springer Nature Customer Service Center GmbH
Europaplatz 3, 69115 Heidelberg, Germany

Printed by Libri Plureos GmbH
in Hamburg, Germany